Digital Development in Korea

Digital Development in Korea explores the central role of digital information and communication technology in South Korea. Analyzing the role of ICT in green growth and sustainability, this new edition also demonstrates how concerns over public safety and the Olympic Games are shaping next generation digital networks.

Presenting a network-centric perspective to contextualize digital development politically, economically and socially, as well as in relation to globalization, urbanization and sustainability, this book builds on firsthand experience to explain the formulation and implementation of key policy decisions. It describes the revolutionary changes of the 1980s, including privatization and color television and the thorough restructuring that created a telecommunications sector. It then goes on to explore the roles of government leadership, international development and education in affecting the diffusion of broadband mobile communication, before weighing up the positive and negative aspects of Korea's vibrant new digital media.

Seeking to identify aspects of the Korean experience from which developing countries around the world could benefit, this book will be of interest to students, scholars and policymakers interested in communications technologies, Korean studies and developmental studies.

Myung Oh is a former Deputy Prime Minister of Korea, who additionally held four ministerial positions under four different administrations, beginning with service as Minister of Communications in the 1980s. The founder and honorary president of SUNY Korea, he holds a Ph.D. in electrical engineering from Stony Brook University, USA.

James F. Larson is a Leading Professor at SUNY Korea where he previously served as Provost and Chair of the Department of Technology and Society. He holds a Ph.D. in communication from Stanford University, USA.

Routledge Advances in Korean Studies

For more information about this series, please visit: www.routledge.com/asianstudies/series/SE0505

Digital Development in Korea

Lessons for a Sustainable World

Second Edition

Myung Oh and James F. Larson

Routledge
Taylor & Francis Group

LONDON AND NEW YORK

Second edition published 2020
by Routledge
2 Park Square, Milton Park, Abingdon, Oxon, OX14 4RN

and by Routledge
52 Vanderbilt Avenue, New York, NY 10017

Routledge is an imprint of the Taylor & Francis Group, an informa business

First edition published by Routledge 2011

British Library Cataloguing-in-Publication Data
A catalogue record for this book is available from the British Library

Library of Congress Cataloging-in-Publication Data
Names: O, Myung, author. | Larson, James F., author.
Title: Digital development in Korea : lessons for a sustainable world /
    Myung Oh and James F. Larson.
Description: Second edition. | Abingdon, Oxon ; New York, NY :
    Routledge, 2020. | Series: Routledge advances in Korean studies |
    Includes bibliographical references and index. |
Identifiers: LCCN 2019013053 (print) | LCCN 2019017317 (ebook) |
    ISBN 9780429022111 (Ebook) | ISBN 9780429666698 (Adobe Reader) |
    ISBN 9780429663970 (ePub) | ISBN 9780429661259 (Mobipocket
    Encrypted) | ISBN 9780367076917 (hbk) | ISBN 9780429022111 (ebk)
Subjects: LCSH: Information society—Korea (South) | Digital
    communications—Korea (South) | Sustainable development—
    Korea (South)
Classification: LCC HM851 (ebook) | LCC HM851 .O3 2020 (print) |
    DDC 303.48/33095195—dc23
LC record available at https://lccn.loc.gov/2019013053

ISBN: 978-0-367-07691-7 (hbk)
ISBN: 978-0-429-02211-1 (ebk)

Typeset in Times New Roman
by Apex CoVantage, LLC

To all those who contributed to Korea's ICT development and also to the young people and leaders who are striving for economic growth in developing countries around the world.

Contents

Figures

Tables

Preface to the second edition

Given the pace of change in digital networks and related technology, the authors anticipated the eventual need for a revision of their 2011 book. This book represents our effort to meet that need, and it too will undoubtedly require an update some years hence.

This edition contains several major changes from the first edition, which was published in 2011. First, most references to quantitative data and longitudinal trends are updated to include the most recent data available.

Second, the chapter structure was modified so that the main topics covered in Chapters 1 through 9 of the 2011 edition are treated in Chapters 1 through 6 of this edition. This allowed for the inclusion of three entirely new chapters. Chapter 7 introduces the topic of green growth and sustainable development. Chapter 8 examines the role of digital networks in relation to the Olympics, given that Korea hosted the Pyeongchang Winter Games in February of 2018. Chapter 9 looks at next generation networks with particular attention to the convergence of 5G and public safety LTE networks.

Third, there is a stronger thematic emphasis in this new edition on what we call "network-centric digital development" in Korea. This begins with an extended treatment in Chapter 2 of the epochal "telecommunications revolution of the 1980s" during which the legal, institutional and policy foundations for Korea's subsequent digital development were laid. Chapters 8 and 9 stress the contemporary importance of networks in Korea's digital development. Just as Dr. Oh Myung and Dr. Kim Jae Ik saw telecommunications as central to the growth of Korea's electronics sector in 1980, next generation networks today are central to the development of augmented reality, virtual reality, artificial intelligence, blockchain and related digital technologies.

For those who wonder how this book came about, the preface to the first edition tells the story.

Preface to the first edition

Although the authors share a strong interest in the role of information and communications technology (ICT) in national development, the story of how this book came to be written is more involved, and a bit serendipitous. It was Choe Chongho, a Yonsei University Mass Communication professor who first introduced the authors in 1991. At that time, Dr. Oh Myung served as Chairman of the 1993 Taejon International Expo Organizing Committee. A graduate of the Korean Military Academy and Seoul National University's department of electrical engineering, he had earned a Ph.D. in electrical engineering from the State University of New York at Stony Brook. In the 1980s he had served for seven years and seven months as Vice Minister and then Minister of Communications.

The two authors first met at Dr. Oh's temporary offices at the COEX convention and exhibition complex in southern Seoul, as he was heavily involved with planning for the forthcoming Taejon International Expo. At that meeting, Dr. Oh sounded out Larson about his possible interest in conducting research on Korean telecommunications. The meeting went well and within months Dr. Oh had helped Larson secure funding from Dacom Corporation to support the research that culminated with publication of his 1995 book, *The Telecommunications Revolution in Korea* (Oxford University Press). That book was the first English language scholarly monograph devoted to the 1980s, a critical period in revitalizing Korea's electronics sector and jump-starting its digital development.

Over the next decade or more the author's paths diverged in interesting ways. Following the Taejon International Expo, Dr. Oh served very briefly as Commissioner of the Korea Baseball Organization and was then called upon once again for government service as Minister of Transportation and then Minister of Construction and Transportation from 1993–1995. From 1996 through 2001 his career took another turn as he became President and then Chairman of the Dong-a Daily Newspaper. After that he served as President of Ajou University in Suwon from March 2002 through December 2003, at which point he was again summoned to serve his country as Minister of Science and Technology, a post that saw him designated as Deputy Prime Minister and Minister of Science and Technology from October 2004 through February 2006.

While Dr. Oh continued his career as a government leader, Larson worked as an administrator with the Fulbright Commission in Seoul, with responsibilities for technology and web-based services. When they met again in the spring of 2007, Dr. Oh was some months into his new position as President of Konkuk University, and both were acutely aware of the need for an updated or completely new book on Korea's digital development. Events since the early 1990s, especially in broadband internet and mobile communication, had carried Korea well beyond the subject matter of Larson's 1995 book. Dr. Oh found himself dedicating an increasing amount of time visiting developing countries who were seeking Korea's advice on how to best use ICT for development.

The authors met several times to discuss common concerns. However, their interests were somewhat different. Dr. Oh was considering writing a memoir and enlisted Larson's help, sharing with him a number of his published and unpublished Korean language manuscripts. As he translated most of those documents into English, Larson was all the while thinking of writing a second, thoroughly updated edition of *The Telecommunications Revolution in Korea*.

It was in September of 2008 that the two met and had a critical meeting of the minds that led to this book. At that meeting Larson related to Oh that the publisher of his 1995 book was not interested in considering a proposal for an updated second edition, mainly because editors thought the market in the U.S. and Europe would be too small. Dr. Oh immediately replied, "But people in the developing countries are very interested in this topic!" He had just returned from a visit to two Latin American countries that were, indeed, very interested in learning from Korea's experience. Larson then asked whether Dr. Oh would like to co-author a book, to which the reply was "Of course, why not?" In fact, neither of the authors had any reason to hesitate on the matter. Dr. Oh was increasingly in demand to consult with leaders of developing nations on behalf of the Korean government. Larson's interest in communication for development dated from his Peace Corps experience in Korea and graduate work at Stanford in the 1970s. It was apparent that the two shared a strong interest in communication for development, including Korea's experience and the possible lessons it might hold for other developing countries. Agreement to co-author a book meant that they could combine their perspectives and experience to hopefully produce something unique and stronger than a book authored by either one of them alone.

Although Dr. Oh's government career began in the telecommunications sector and he is widely acknowledge as a leader of the 1980s telecommunications revolution, it grew from those roots to encompass several other key industries and technologies that were benefiting from the information revolution and converging with ICT. Notably, Dr. Oh played a central role in fostering the biotechnology industry and in the development of space technology while serving as Deputy Prime Minister and Minister of Science and Technology. During his tenure Korea enacted the space development act and sent a Korean into space for the first time in history. The nation also built a satellite for the first time with indigenous technology and launched it from Korean land into orbital space.

Acknowledgments

Although this book focuses on digital networks and technologies, it is to individuals and human networks that we owe our greatest debt of gratitude. The authors' interpersonal networks have grown so much over time that it is impossible to acknowledge all of the individuals, and even in naming organizations and groups of people we risk omission of some.

Internationally, the work of researchers from such organizations as the World Bank, the ITU and the OECD has provided much of the raw material for our analysis. We acknowledge their efforts, especially the continued attempt to better measure the information society through the specific contributions of ICT to development. From 2016 through 2017 Larson served as a member of an ITU expert group and authored Chapter 7, "The Future of ICT-driven Education for Sustainable Development" in the book produced by the group entitled *ICT-centric Economic Growth, Innovation and Job Creation* (2017).

In Korea, the authors gratefully acknowledge assistance from the leadership and staff of several organizations. They include the National Information Society Agency (NIA), the Korea Communications Commission (KCC) and Korea Telecom. We offer special thanks to the leadership of the Electronics and Telecommunications Research Institute (ETRI) and to the Korea Information Society Development Institute (KISDI).

Professor Park Jaemin of the Graduate School of Technology Management at Konkuk University helped greatly with a final review of the manuscript. Also, the authors would like to acknowledge the numerous colleagues in government, industry and academia whose work contributed in one way or another to this book. Our intellectual and personal debt to them will be apparent to those who read this book. To all of you, we gratefully acknowledge your assistance in making this study possible.

Abbreviations

ADSL	Asymmetric Digital Subscriber Line
AFKN	Armed Forces Korea Network
AI	Artificial Intelligence
AMOLED	Active Matrix Organic Light Emitting Diode
AMPS	Advanced Mobile Phone System
AR	Augmented Reality
ARPANET	Advanced Research Projects Agency Network
ASIC	Application-Specific Integrated Circuit
ATM	Asynchronous Transfer Mode
BEM	Building Energy Management
BERD	Business Enterprise R&D Expenditure
CDMA	Code Division Multiple Access
DAC	Development Assistance Committee, a committee of the OECD
DACOM	Data Communications Corporation of Korea
DARPA	Defense Advanced Research Projects Agency, a U.S. government agency
DMB	Digital Multimedia Broadcasting
DMZ	Demilitarized Zone, a zone established at the 38th parallel when the armistice ended the Korean War, separating North and South Korea
DRAM	Dynamic Random Access Memory
EIP	Eco-industrial Park
EMS	Emergency Medical Services
EPB	Economic Planning Board
ETRI	Electronics and Telecommunications Research Institute
FDI	Foreign Direct Investment
GDP	Gross Domestic Product
GHGE	Greenhouse Gas Emissions
GIONS	Games Information Online Network
GNI	Gross National Income
GSM	Global System for Mobile Communication
HCI	Heavy and Chemical Industries, a sector of the Korean economy given priority during the latter years of Park Chung Hee's presidency
IC	Integrated Circuit

ICR	Institute for Communication Research
ICT	Information and Communications Technology
IDI	ICT Development Index, an index developed by the ITU to measure the relative diffusion and use of information and communication technologies in countries around the world
IFEZ	Incheon Free Economic Zone
IMF	International Monetary Fund
IOC	International Olympic Committee
IPF	Information Promotion Fund
IPTV	Internet Protocol TV
ISIC	International Standard Industrial Classification
ISP	Internet Service Provider
ISTK	Korea Research Council for Industrial Science and Technology
IMF	International Monetary Fund
IT	Information Technology, the branch of engineering dealing with the use of computers to retrieve, store and transmit information
ITU	International Telecommunications Union
KADO	Korea Agency for Digital Opportunity and Promotion
KAIS	Korea Advanced Institute of Science, predecessor to KAIST
KAIST	Korea Advanced Institute of Science and Technology
KCC	Korea Communications Commission
KDI	Korea Development Institute
KEPCO	Korea Electric Power Corporation
KETEP	Korea Institute of Energy Technology Evaluation and Planning
KETRI	Korea Electronics and Telecommunications Research Institute
KFTC	Korea Fair Trade Commission
KIAT	Korea Institute for the Advancement of Technology
KIET	Korea Institute for Industrial Economics and Trade
KII	Korea Information Infrastructure
KISDI	Korea Information Society Development Institute
KISI	Korea Internet and Security Agency
KIST	Korea Institute of Science and Technology
KISTEP	Korea Institute of Science and Technology Evaluation and Planning
KMT	Korea Mobile Telecom
KOICA	Korea International Cooperation Agency
KOSEF	Korea Science and Engineering Foundation
KPX	Korea Power Exchange
KRCF	Korea Research Council of Fundamental Science and Technology
KSGA	Korea Smart Grid Association
KSGSF	Korea Smart Grid Standardization Forum
KT	Korea Telecom
KTA	Korea Telecommunications Authority
KTX	Korea Train Express, Korea's high-speed rail system
LCD	Liquid Crystal Display
LED	Light Emitting Diode

LSI	Large Scale Integration
LTE	Long Term Evolution, a 4G mobile communications standard
MCPTT	Mission-Critical Push-To-Talk
MIC	Ministry of Information and Communication
MITI	Ministry of International Trade and Industry, a Japanese ministry
MKE	Ministry of Knowledge Economy
MMOG	Massive Multiplayer Online Game
MOC	Ministry of Communications
MOE	Ministry of Education and Human Resources
MOOC	Massive Open Online Course
MOPAS	Ministry of Public Administration and Security
MOSF	Ministry of Strategy and Finance
MOST	Ministry of Science and Technology
MOTIE	Ministry of Trade, Industry and Energy
MSIP	Ministry of Science, ICT and Future Planning
MSM	Mobile Station Modem
NAND	Stands for Not-And, a type of flash memory
NCA	National Computerization Agency
NGO	Non-governmental Organization
NIA	National Information Society Agency
NRF	National Research Foundation
NSC	National Security Council
NSTC	National Science and Technology Council
OECD	Organization for Economic Cooperation and Development
OEM	Original Equipment Manufacturer
OPC	Oriental Precision Company
OSTI	Office of Science and Technology Innovation
PCGG	Presidential Committee on Green Growth
POSCO	Pohang Iron and Steel Company
POTS	Plain Old Telephone Service
PPP	Public-Private Partnership
PS-LTE	Public Safety LTE, refers to new public safety networks built using the Long Term Evolution 4G standard
PSTN	Public Switched Telephone Network
SAIT	Samsung Advanced Institute of Technology
SDG	Sustainable Development Goal
SGS	Smart Grid Station
SITC	Standard International Trade Classification
SMS	Short Message Service
SNU	Seoul National University
TDMA	Time Division Multiple Access
TDX	Time Division Exchange, the name given to South Korea's first electronic switching system
TETRA	Terrestrial Trunked Radio
TOP	The Olympic Program, a program for global Olympic sponsors

U.S. AID	Agency for International Development, a U.S. government foreign aid agency
VLSI	Very Large Scale Integration
VOIP	Voice Over Internet Protocol
VR	Virtual Reality
WiBRO	Wireless Broadband, a variant of mobile WIMAX technology developed in Korea and accepted as an international standard
WINS	Wide Information Network System
WIPI	Wireless Internet Platform for Interoperability, a mobile communications standard used in Korea for some years
WTO	World Trade Organization

Introduction

The foundation for South Korea's remarkable digital development over the past four decades was laid in the 1980s. That reality led many who are familiar with Korea's situation to refer to the "telecommunications revolution of the 1980s." As this book will argue, the breakthroughs Korea made during that decade, most especially in electronic switching and the semiconductor industry, were a necessary pre-condition for many ensuing developments. This much can be inferred from the basic cumulative characteristic of innovation in information and communication technologies. In economic terms, information is both an input and an output of its own production process. This is referred to by economists as the "on the shoulders of giants" effect. Absent the 1980s success in research and development along with key policies and the development of citizen awareness, South Korea would be at a different stage of development today.

Our perspective on Korea's digital development is shaped not only by theory, but also by practice and experience. This book's first author, Dr. Oh Myung,[1] was a prominent leader of the telecommunications sector during the 1980s. He led the shaping of major policies, the influence of which was felt through the 1990s and beyond. Those major policies and changes are highlighted throughout this introduction.

Reorganization of the telecommunications management system

The first major policy change was a fundamental and historic restructuring of the telecommunications management system. Until 1980, everything in South Korea's telecoms sector was handled through the Ministry of Communications (MOC), a government monopoly. The new framework introduced by Dr. Oh and his colleagues consisted of a policymaking arm, a business operation arm and research and development institutes. These three areas comprised a specialized system in which each organization was assigned specific functions.

The Telecommunications Policy Office, established on January 1, 1982, was given responsibility for formulating and promoting telecommunications policies in the public interest. It supervised and supported the public telecommunications manufacturing and construction industries. In January of 1984 the

Ministry restructured the Telecommunications Policy Office into four divisions, responsible for Telecommunications Planning, Telecommunications Promotion, Telecommunications Management and Information/Communications. The purpose of this reorganization was to strengthen the information and communication sectors.

The business operations in Korean telecommunications were bolstered by the establishment of the Korea Telecommunications Authority (KTA) and DACOM Corporation. KTA was established on January 1, 1982, to administer telecommunications business operations and pursue profitability based on the public interest. The government was the sole investor, contributing 2.5 trillion won to establish the Authority. At the time, a total of 35,225 employees from 153 divisions moved from the MOC to KTA, the largest reshuffling of personnel from one organization to another in the history of the Korean government. However, there were few problems during the transition, and it set an example for the subsequent establishment of the Korea Tobacco Monopoly Corporation.

In an effort to boost the sagging data communications sector, the MOC decided that a private company needed to administer business operations in that field. Consequently, on March 29, 1982, it established the Data Communications Corporation of Korea (DACOM) under a business promotion committee on which Dr. Oh Myung served as a commissioner, in his capacity as Vice Minister of Communications. In April of the same year, the first value-added telecommunications service was licensed to provide public data transmission, processing and databank services. In 1984, the Korea Mobile Telecommunications Corporation was established to oversee mobile and paging services. A year later, the Korea Port Telecommunications Corporation was formed to handle wired and wireless services relating to Korea's ports.

The formation of research and development institutes for the telecommunications sector also began during the 1980s. On March 26, 1985, the Telecommunications Research Institute of Korea was merged with the Electronics and Telecommunications Research Institute (ETRI). In February of the same year, the Institute for Communication Research (ICR) was established as a think tank for policymaking. In January of 1988 it was renamed the Korea Information Society Development Institute (KISDI).

Resolution of the telephone backlog and nationwide telephone automation

A backlog in fulfilling requests for telephone service, as described in more detail in Chapter 2, was one of the most serious problems affecting Korean society during the 1970s. However, starting in 1981, the government installed an additional 1 million lines and made an average annual investment of 1 trillion won. As a result, the number of telephone lines exceeded 10 million as of October 1987, making Korea the tenth ranked country in the world in terms of the number of lines installed. This made it possible for every Korean household to be equipped with a telephone.

In 1984 Korea became the first nation in the world to complete construction of a digital toll switching network. By 1987 the chronic telephone service backlog had been completely eliminated and an immediate installation system introduced.

The year 1987 was also significant for completion of the nationwide telephone automation system. By June of that year Korea had an automated long-distance calling service, promotion of subscription in rural areas and automated telephone service on islands. Taken together, these projects meant realization of the goal of a nationwide automated calling system. It allowed callers to receive direct inter-national call service to 100 countries worldwide as well as immediate local and toll-call services. Service was provided to a total of 25,000 villages, each with at least 10 households, in mountainous areas and to 500 island villages as well.

Development of the TDX electronic switching system

In retrospect, the domestic development of the TDX electronic switching system is regarded as the most outstanding achievement in Korea's telecommunications sector over the years. Even Daniel Bell expressed his astonishment at Korea's early development of an electronic switching system. Korea became the tenth nation in the world to accomplish the proprietary development of an electronic switching system and the sixth nation to export such a system to foreign countries. The development of TDX not only triggered rapid progress in Korean information and telecommunications technology, but also substantially reduced the consider-able expense associated with importing foreign-made equipment.

Deregulating the supply of terminals

Until the 1970s, the Ministry of Communications itself was the only supplier of household telephone terminals as well as the only lender of telephone sets. This meant that subscribers did not have the option of owning a telephone or choos-ing which model of telephone they received. However, in January of 1981 the MOC introduced a policy that gave subscribers the opportunity to choose from a variety of options. On September 1, 1985, the MOC transferred the ownership of 2.27 million telephone sets from the KTA to telephone subscribers themselves. Consequently, manufacturers began to produce telephone sets with diverse func-tions and various designs, giving buyers a wider range of choices. At the same time, terminals for mobile telephones and telex machines were made available to the public, and complete liberalization of terminal distribution became a reality.

Deregulation of the public switched telephone network (PSTN) and radio wave use

In keeping with worldwide trends in digital communications, facsimile machines, modems and personal computers began to find their way into offices and homes throughout Korea during the early 1980s. Prior to 1982 these devices could not be connected to the PSTN. Instead, they were connected to lines specifically designated for such devices. However, in March of 1983 the PSTN became available to the public and provided non-voice telecommunication services for facsimiles, modems and computers. It is worthwhile noting that this move preceded such deregulation of the PSTN in most European countries.

Until the 1970s, private use of radio-wave frequencies had been restricted for national security reasons. In December of 1982 the MOC introduced a paging service and, a little later, a mobile communications service, initially only in metropolitan areas. The use of cordless phones was permitted from September of 1983. Also, integrated management of broadcasting networks was introduced at the end of 1986, improving their operation and increasing the service coverage rate from 66 percent in 1980 to 94 percent in 1987.

Color television and financing the telecommunications revolution

In addition to the above policies, there were two other related issues dealt with by the Korean government during the 1980s: the introduction of color television broadcasting and the securing of financial resources for investment in telecommunications.

The first television broadcast in Korea took place on June 1, 1956. It was not until October 1980 that, despite fierce opposition, the government decided to start color television broadcasting. The first color telecast took place on December 1 of that year. This was 29 years behind the United States and 20 years behind Japan.

The development of Korea's telecommunications sector required a tremendous investment. During the formulation of basic plans for telecommunications development in the early 1980s two measures were taken to guarantee a stable investment in telecommunications infrastructure. First, the 1981 revision of the Law on the Expansion of Information Facilities, first enacted in 1979, encouraged subscribers to purchase a fixed number of telephone bonds at the time of telephone installation. In Seoul, this cost a new subscriber 200,000 won. Second, the relatively low telephone charges at the time were adjusted to a more reasonable level. The local call rate increased from 8 won to 12 won in January 1980. It rose to 15 won in June 1981 and to 20 won in December of the same year. During the severely inflationary environment of the times, this measure was thought necessary to secure telecommunications funding.

Dr. Oh Myung as evangelist for the information society

During the revolutionary decade of the 1980s Dr. Oh Myung established his credentials as a tireless proponent of the information society. His underlying philosophy included a deep commitment to creation of an equitable information society in which all citizens had access to the benefits of informatization and where everyone could enjoy a quality life regardless of where they lived. He took advantage of every opportunity to speak on the topic of the information society and engage in dialogue on the topic with government officials, professors, students and citizens in many organizations. His work as an evangelist for IT innovation continued later in his career and extended to developing countries around the world including Paraguay, Columbia and Rwanda. International organizations such as the International Telecommunications Union (ITU) also frequently called upon him for advice based on his leadership in Korea's ICT-driven development.

Chapter 2 will go into further detail on Korea's "telecommunications revolution of the 1980s" and the start of its prodigious efforts to build an equitable information society. The following chapters all help to explain how the nation's network-centric digital development grew from those revolutionary developments in the 1980s and the prospects for the future.

Note

1 Since Dr. Oh Myung is both a well-known public figure and the first author of this book, he will be referred to throughout by name, rather than as "this book's author." Also, as with other Korean names referred to in the book, we will place the family name, in his case Oh, first followed by his given name, Myung. Note that Dr. Oh's given name is a single syllable whereas most Korean given names consist of two syllables.

1 Digital development as Korea's destiny

Placing the "Miracle on the Han" in context

Korea is a mountainous peninsula, with mountains occupying over 70 percent of the land area. The mountain ridges and the rivers that flow through adjacent valleys form a dense network, clearly visible from space that has shaped Korean culture and patterns of human communication from time immemorial. As long ago as the Chosun Dynasty, smoke and fire beacons were used to speed communication throughout the nation's mountainous terrain. Mountains are not just an indelible part of Korea's physical world, but of her mentality and consciousness as well. The influence of mountains is deeply embedded in the emotions, knowledge, beliefs and values of Korea.[1] For example, mountain ridges play a central role in one of Korea's treasured myths as expressed in the "*Arirang*" folk song.

Today a new, denser set of advanced, digital communication networks weaves its way throughout the southern half of the Korean Peninsula. Less visible to the human eye, these fiber optic, mobile and satellite networks have propelled South Korea from a follower in electronics and telecommunications to a world leader in the field. As a 2003 report by the ITU (International Telecommunications Union) put it, "Korea is the leading example of a country rising from a low level of ICT access to one of the highest in the world."[2] An extensive study by the World Bank of Korea's emergence as a knowledge economy came to essentially the same conclusion.[3] After the turn of the millennium Korea solidified its position as an ICT leader. As shown in Table 1.1, Korea maintained its position at or near the top of world rankings on the ITU's ICT development index (IDI), which is composed of sub-indices measuring access, use and skills.[4] It ranked number one in five of the seven years for which the ITU reported data from 2010 through 2017.[5]

The Republic of Korea's development story, rising as it did from the ashes of the Korean War to its present status as one of the world's most economically advanced nations, is the source of its appellation the "Miracle on the Han." Furthermore, the electronics industries became the main driver of Korea's economic growth beginning in the late 1980s and Korea's burgeoning ICT sector was the main engine of economic growth in the decades since. During the last half of the 20th century Korea alone, among developing nations, rose from the status of an aid recipient to that of an aid donor. In 2010 it officially became a member of the OECD's Development Assistance Committee (DAC).[6] No other country in the world has achieved such success against such odds.

Table 1.1 ICT Development Index (IDI) for top-ranked nations, 2010–2017

	1st	2nd	3rd	4th	5th
2010	Korea	Sweden	Denmark	Iceland	Finland
	8.45	8.21	8.01	7.96	7.89
2011	Korea	Sweden	Denmark	Iceland	Finland
	8.51	8.41	8.18	8.12	7.99
2012	Korea	Sweden	Iceland	Denmark	Finland
	8.57	8.45	8.36	8.35	8.24
2013	Denmark	Korea	Sweden	Iceland	UK
	8.86	8.85	8.67	8.64	8.5
2014	n.a.	n.a.	n.a.	n.a.	n.a.
2015	Korea	Denmark	Iceland	UK	Sweden
	8.93	8.88	8.86	8.75	8.67
2016	Korea	Iceland	Denmark	Switzerland	UK
	8.8	8.78	8.68	8.66	8.53
2017	Iceland	Korea	Switzerland	Denmark	UK
	8.98	8.85	8.74	8.71	8.65

Source: ITU, *Measuring the Information Society* reports.

While miracles by their very nature defy explanations, the "Miracle on the Han" can be better understood if placed in a sharper and clearer context. The iconic photograph in Figure 1.1 of the Korean Peninsula at night symbolically suggests the importance of such context. Taken from the International Space Station in January of 2014,[7] it shows at a glance the perspective added by distance. Several elements of the photograph stand out. First, there is the stark contrast between the darkness in North Korea and the light throughout most of South Korea, reflecting the South's rapid urbanization and socioeconomic development, which was not shared by the North. Second, the photograph shows the concentration of light and population in South Korea's major urban areas, led by the national capital metropolitan area around Seoul, and including Busan, Daegu, Daejon and other major cities. Third, the demilitarized zone, which curves along the 38th parallel, is clearly delineated by lighting along its southern border. This military division represents the cease-fire agreement rather than an official end to the Korean War. The one small dot of light north of the DMZ beyond the northern edge of the Seoul metropolitan area is Kaesong, site of the industrial complex jointly operated by North and South Korea. Finally, Korea's mountain valleys mentioned at the outset are shown in this photograph as lines of light along transportation routes that still to this day follow along rivers and streams, connecting cities and villages. Some of the most mountainous areas in the northeastern Gangwon province and further down the east coast stand out from the big urban areas. In short, even this snapshot from space suggests the sort of context needed to fully understand Korea's miraculously rapid socioeconomic development over recent decades.

The primary focus of this book is on the role of digital communication networks and related technologies in South Korea's development since 1980. Given Korea's mountainous terrain, its relative lack of natural resources and the country's utter

Figure 1.1 The Koreas at night
Source: NASA Earth Observatory

devastation in the wake of the Korean War, it was an unlikely candidate to become a leader in digital technologies. As of 1980 Korea was still a poor developing country, unable to provide even basic, plain old telephone service (POTS) to most of its citizens. In such circumstances, digital development represented a course that could be pursued through education and sustained human effort. In retrospect, it seems that digital development was the nation's destiny.

Within Korea, especially among the technocrats and engineers who planned and built the nation's networks, the introduction of digital networks came to be called the "1980s telecommunications revolution." Indeed, Dr. Oh Myung is widely recognized as the principal architect and godfather of that revolution. However, the story of Korea's digital development is still only vaguely understood by many international observers, recalling Isaacs' observation in 1958 that "Vagueness about Asia has been until now the natural condition even of the educated American."[8] Isaacs was explicitly concerned with the relationship of image holding to policymaking and based his findings on panel interviews with nationally prominent representatives from academia, the media, government and business.

In today's digital media environment, vagueness and misperceptions of Korea persist and, as in Isaacs' era, they can only detract from sound and effective policymaking. One major reason for such vagueness is the lack of historical, cultural, political or economic context. The historical or temporal relationship of digital innovation in Korea to that taking place in companies and countries around the world assumes importance for two reasons. First, Korea reaped advantages because it introduced digital networks approximately at the same time they were first being tested and installed in other parts of the world. Second, one must consider the cumulative nature of the technologies that go into the construction of advanced digital networks. Existing knowledge and expertise is needed to build the next generation of networks, illustrating what economists call the "on the shoulders of giants" effect.

In cultural terms, there is the sheer difficulty for non-Koreans of mastering Korea's language and culture. The Foreign Service Institute in the U.S. Department of State ranks Korean along with Chinese (both Mandarin and Cantonese) and Japanese as an exceptionally difficult and time consuming language for native English speakers to learn.[9] Consequently, some published studies show a tendency to conflate characteristics of Korea with those of its larger neighbors, China and Japan.

To illustrate the problem, consider two glaring errors in the otherwise excellent 2003 study conducted by experts for the International Telecommunications Union (ITU). In the opening chapter the authors erroneously claim that the Korean alphabet, *hangul*, weighed against the country's ICT development because it used a pictographic font that is not ideally suited to computerization.[10] Exactly the opposite is true as *hangul* consists of consonants and vowels, not pictographs, and is ideally suited to computerization and various types of keyboard input. *Hangul* not only allowed the rapid achievement of near-universal literacy in Korea during the latter half of the 20th century, but it was also a significant factor driving ICT literacy.[11]

The ITU study also contains another major error most likely deriving from a lag in international reporting of data. It presents a bar graph depicting the size of the waiting list for telephone service by year from 1982 through 1992.[12] Unfortunately, the graph indicates that the backlog (waiting list) in provision of telephone service persisted through 1988, when in fact it had been eliminated almost two years earlier, with completion of the nationwide public switched telephone network in June of 1987.[13]

Another example of misperceptions can be found in Fransman's book about the new ICT ecosystem.[14] In it he stresses the importance of each country prioritizing its goals and objectives and then benchmarking performance against the global leaders and a group of comparable countries. Using the example of broadband he suggests that international benchmarking "reveals that Japan is significantly ahead of the rest of the world."[15] Unfortunately, the empirical evidence contradicts his argument that Japan has the fastest internet download speeds in the world. As illustrated by Figure 1.2 for over a decade since 2007 Korea's broadband networks have consistently offered the fastest average download times in the world, exceeding those of Japan, Hong Kong and the U.S.

Based on data collected in the first quarter of 2014, the OECD Broadband Portal reported that measurements by M-Lab and Ookla also showed Korea with a significant lead in download speeds over Japan.[16]

A major purpose of this book is to help policymakers, industry leaders, academics and citizens better contextualize South Korea's network-centric, ICT-driven development. It focuses on the origins of the nation's digital networks,

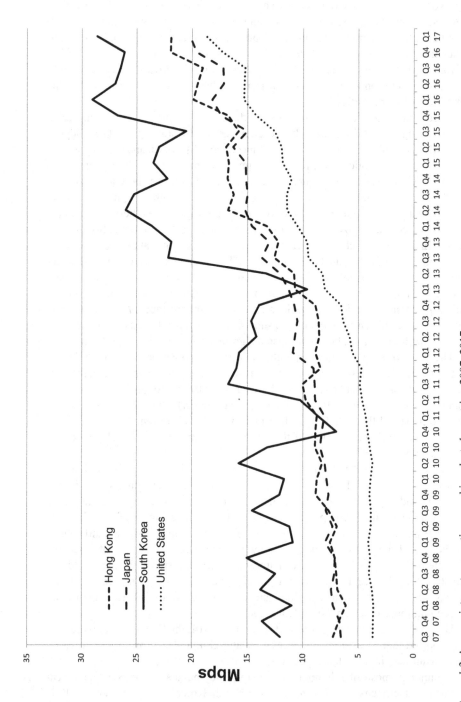

Figure 1.2 Average internet connection speed in selected countries, 2007–2017

Source: Akamai State of the Internet Reports

their expansion in recent decades, their contribution to socioeconomic development and prospects for next generation networks. A main concern throughout is with policy issues bearing directly on digitization of networks, network architecture and network technologies, along with the historical, cultural, political and economic context in which these policies were considered, drafted and implemented.[17] This book addresses the following questions:

- How could a resource-poor nation whose industrial development in the 1960s and 1970s focused on heavy manufacturing industries make the transition to digital network-driven development in the 1980s?
- How and why did the ICT sector emerge as the main engine of Korean development, and what policies were implemented to accomplish this?
- What was the nature of government-led telecommunications policy, including the changing nature of the government-industry relationship over time?
- When did Korea's transformation originate relative to global developments, and what are the prospects for Korean involvement in shaping next generation networks?
- What sorts of funding mechanisms did Korea use to finance its digital development, and how were these implemented?
- What role did education and citizen awareness play in Korea's digital development?
- What lessons, both successes and failures, might other countries take away from a careful examination of the Korean experience?

The digital disruption, globally and in Korea

This book examines both the digital transformation in Korea and its place in that same revolution on a global scale. Just as globalization and international developments helped to shape the Korean ICT sector, Korea's digital development made it an influential participant in the global dialogue about ICT sector technologies, trade, policies, standards and related matters.

Globally, the digital network revolution originated with three important mid-20th century developments. First came the invention of the transistor by Shockley and colleagues at Bell Labs in 1947 for which they received a Nobel Prize.[18] It opened the way for subsequent development of the semiconductor industry. Second was the publication in 1948 of Claude Shannon's mathematical theory of communication, which launched information theory and allowed the digitization of all kinds of information, including text, voice, music and video.[19] These two innovations enabled engineers to build completely digital switches for installation in telephone networks. The digitization of networks in turn led to growth and development of the internet, the largest engineered system in human history, with hundreds of millions of connected computers, links and switches and billions of users connecting through a growing array of devices.[20] Along with energy and transportation, digital networks are now widely recognized as a key infrastructure for the 21st century.

The digitization of networks had sweeping economic, political and social impact, perhaps nowhere more so than in South Korea. As explained by Benkler, the economic ecology of the 20th century industrial mass media era changed to one in which the basic output in most advanced economies is human meaning and communication. In this new ecosystem the only physical capital needed to express and communicate human meaning is the connected personal computer.[21]

In economic terms, information itself has several unique characteristics. First, it is nonrival, meaning that its consumption by one person does not rule out its consumption by others. Second, it has a marginal cost near zero. Third, information is both an input and an output of its own production process. Simply put, this means that one must use existing information to produce new information.[22] These characteristics of information apply to the design and implementation of new digital network technologies, a process that involves rapid and cumulative changes in both hardware and software. Consequently, the starting point in digital development, which differs country by country, is so important.

The digital network revolution has already produced exponential increases in the world's technological capacity to store, communicate and compute information. Hilbert and Lopez documented this growth during the period from 1986 to 2007 by tracking 60 analog and digital technologies over this period. Their research showed that in 1986 only 20 percent of information transmitted by telecommunications networks was analog. By 1993 that had increased to 68 percent. In 2000 it rose to 98 percent, and by 2007 99.9 percent, virtually all, information transmitted on telecommunications networks was digital. Increases in memory and computing capacity showed a similar pattern. By 2007 digital memory constituted 94 percent of the world's technological memory.[23]

The diffusion of digital networks to countries around the world was an uneven process, proceeding more quickly in advanced developed economies than in the poor, developing ones, leading to scholarly concern with the concept of "digital divide."[24] In a striking exception to this global pattern, South Korea moved decisively in the early 1980s to digitize its telephone network, completing a fully digital public switched telephone network (PSTN) in June of 1987. Arguably Korea's most important network innovation of the 1980s was the successful TDX switching project, which facilitated completion of the nationwide PSTN. That gave Korea one of the technologically most advanced such nationwide networks at the time, surpassing Japan and other advanced nations.

Completion of the PSTN preceded similar efforts by most other countries and coincided with the introduction of digital switching by the world's leading telecommunications firms and in advanced economies. Such fortunate timing helped Korea to harness the power of digital networks for national socioeconomic development and to do so more successfully than any other developing country to date. Consequently, as later chapters will discuss in detail, the development and implementation of next generation networks including 5G and the Internet of Things (IoT) are further along in Korea than in most other nations.

As shown in Figure 1.3, June of 1987 is approximately where Korea's GNI per capita reaches an important inflection point. To the extent that ICT and the new

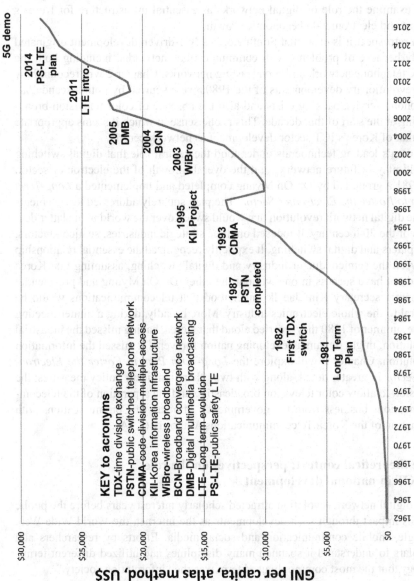

Figure 1.3 Korea's GNI per capita and its major network initiatives

Source: Based on GNI data from the World Bank

digital networks played a central role in Korea's ensuing economic development, this inflection point seems to be no sheer coincidence. The innovation of nation-wide digital networks allowed fax machines, computers, credit card verification units and a plethora of other devices to be connected to the network. Chapter 6 will examine the role of digital networks as essential infrastructure for Korea's export and electronics-led economic growth.

In retrospect it is clear that South Korea's ICT-driven development originated partly because of problems with communications networks, including both the basic telephone network and broadcasting networks. Chapter 2 will recount how the revolutionary developments of the 1980s were spurred by both a scandalous telephone service backlog crisis and also the absence of color television broad-casting at the start of that decade. This is one sense in which it seems appropriate to think of Korea's ICT sector development as network-centric.

Korea's leading technocrats understood the central role that digital switching would play in future networks and the overall growth of the electronics sector. In 1981 a group led by Dr. Oh Myung completed and implemented a *Long-Term Plan to Foster the Electronics Sector*. The plan squarely addressed key elements of the digital network revolution that would sweep over the world in the latter dec-ades of the 20th century. It focused on three strategic industries: semiconductors, computers and digital switching. It explicitly recognized the essential relationship between the semiconductor industry and digital switching, assuming that Korea could not have success in one without the other. Dr. Oh Myung and presidential economic secretary Kim Jae Ik understood that telecommunications would be central to the whole electronics industry. More broadly, during a dinner meeting in the summer of 1980 they worried aloud that Korea, having missed the industrial revolution, might remain a developing nation forever if it missed the information revolution. Chapter 2 will explore the *Long-Term Plan to Foster the Electron-ics Sector* in greater detail, along with two other significant policy measures: the decision to allow color television broadcasting and the separation of the telecom-munications business from the government's Ministry of Communications with formation of the Korea Telecommunications Authority (KTA).

The theoretical context: perspectives on the role of ICT in national development

The digital network revolution attracted scholarly interest years before the public felt its impact through such developments as the internet, the World Wide Web, Google, mobile communication and social media. Efforts by researchers and scholars to understand it spanned many disciplines and utilized different termi-nology, but the most commonly used term was the "information society."

Information society studies

The information society concept represented a vision that began to crystallize in the early post-World War II period.[25] Figure 1.4 shows that the information society

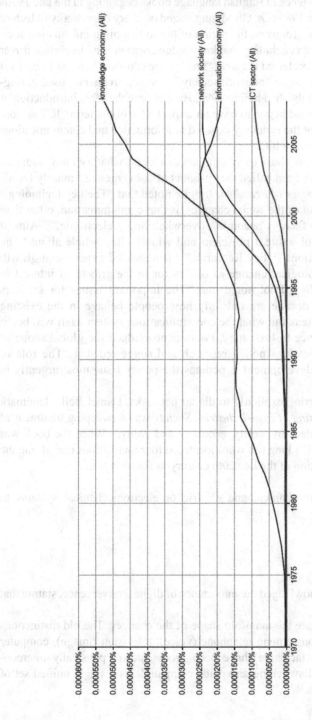

Figure 1.4 Terms used in study of the digital network revolution

Source: Google Books Ngram Viewer.

concept rose to prominence in English language books beginning in the late 1970s and 1980s. During the 1980s Dr. Oh Myung seized on every opportunity to lecture and to establish study groups on the nature of the forthcoming information society. Many economists, concluding that knowledge creation is an important driver of economic growth, preferred the term "knowledge economy," but as Figure 1.4 shows, this term, along with "network society," were not frequently used in English language books published before 1995. By that time the 1992 introduction of the World Wide Web had begun to have an impact. Use of the term "ICT sector" to describe that part of the economy devoted to information and communications technologies (ICT) came a bit later.

Ithiel Pool was one of those early scholars to address what we know today as the ICT sector. He has been called the "prophet of convergence," mainly for his 1983 book, *Technologies of Freedom*. In it he noted that "The key technological change, at the root of the social changes, is that communication, other than conversation face to face, is becoming overwhelmingly electronic."[26] Almost a decade earlier, Pool wrote a prescient and widely cited article about "The Rise of Communications Policy Research."[27] It identified exponential growth in the rate of technological change as one factor in the growth of interest in communications policy. Pool noted that "The important issues for scholars looking at the next decade are not only how people behave in the existing communications system, but what the communications system itself will be."[28] Of particular relevance to this study, Pool acknowledged the global scope of interest in communications policy research and suggested that "The role of communications in development is perhaps the policy issue most urgently in need of study."[29]

Although not referring explicitly to digital networks, Daniel Bell's landmark 1973 book, *The Coming of Post-Industrial Society* was a sweeping treatment of the changes taking place in society, economy and polity.[30] When the book was reprinted in 1999, Bell's long foreword identified four innovations underlying the technological revolution of the late 20th century as follows:

1   the change of mechanical and electric or electromechanical systems to electronics,
2   miniaturization,
3   digitalization,
4   software.

Furthermore, he acknowledged the importance of digital convergence, stating that

> We can already see the manifold shape of the changes. The old distinctions in communications among telephone (voice), television (image), computer (data) and text (facsimile) have been broken down, physically interconnected by digital switching and made compatible as a single unified set of teletransmissions.[31]

The network society perspective

For Manuel Castells, networks are at the heart of today's digital revolution. He prefers the term "network society" to "information society" because information and knowledge have been a central part of societies throughout history. He is very explicit about the novelty of digital networks, stating that "Among information technologies, I include, like everybody else, the *converging set* of technologies in micro-electronics, computing (machines and software), telecommunications/ broadcasting, and optoelectronics."[32] He goes far beyond this, observing, *à la* Marshall McLuhan,[33] that "computers, communication systems and genetic decoding and programming are all amplifiers and extensions of the human mind." Castells' approach is broad enough to encompass all kinds of networks, including financial, military, terrorist, government, NGO and citizens' networks, to name just a few variations.

In an important 2007 study, Jho used the concept of networked governance to explain the rise of Korea's telecommunications industry and its success in the global market.[34] Her study argued for the importance of closer attention to networks and to the strategies employed by political and economic actors facing rapidly changing technology and international circumstances. The role of the state in ICT sector policy, she argued, will differ based on different types of state policies and networks linking the government and the private sector. The emergence of Korea's telecommunications industry in the 1990s, she suggested, represented a transformation into a "centralized network governance." In her words,

> Networked governance is a steering mechanism based on a network that emphasizes coordinated and negotiated patterns of public and private cooperation. It is a centralized network in which the state orchestrates the network of institutions and mechanisms and steers the interests and strategies of the private sector responding to global technology and business networks.[35]

As noted earlier, digital networks spread around the world at an uneven pace. This is characteristic of many technological innovations. According to Rogers, who studied them extensively "An innovation is an idea, practice or object that is perceived as new by an individual or other unit of adoption."[36] He also acknowledged the centrality of communication in the diffusion of innovations. In fact, the second edition of his book was titled *Communication of Innovations: A Cross-cultural Approach* and came out just before he moved to Stanford University to head a major program in communication for development.[37]

Applying Roger's understanding of innovations to the Korean case, we can consider network-centric digital development to be innovative in two main ways, one of them relating to the hardware component of the innovation and the other to the software side. First, the essence of the innovation in terms of hardware involved the manufacturing and assembling of digital switches and other components needed to digitize Korea's networks. The object is the digital network itself, initially the public switched telephone network (PSTN). Second, the software aspect of network-centric innovation involved governance of the nation's emerging ICT sector and the practices associated with widespread adoption and use of networked digital devices. The unit of adoption is the nation, encompassing Korea's government, leading industry groups and broad political support for the innovation from the public.

Developmental state theory

The concepts of networked governance and a "steering network" show the relevance to this study of developmental state theory, which was launched by Chalmers Johnson in his landmark 1982 study of industrial policy in Japan.[38] A major theme of Johnson's study was the extremely complex process of public-private cooperation in postwar Japan that became known as industrial policy. It included the following elements:

- The existence of a small elite and competent state bureaucracy.
- A political system in which the bureaucracy was given sufficient scope to operate effectively.
- Perfection of market-conforming methods of state intervention in the economy.
- A pilot organization like MITI.[39]

Many of the challenges to understanding Korea's ICT-driven transformation over the past four decades are daunting and similar to those faced by Johnson. Just as Japan in the earlier postwar decades had relied on industrial policymaking through strong public-private partnership, South Korea's government and industry leaders worked closely together in a developmental state mode. However, beginning in 1980, in the midst of political, economic and social turmoil following the assassination of President Park Chung Hee the previous fall, Korea's path diverged from that which had been taken by Japan. A group of young, internationally trained technocrats assumed power and changed the direction of Korea's industrial policy during 1980s. Dr. Oh Myung was a leader of that group.

As of 1980 Japan was a world leader, along with the United States, in the semiconductor and consumer electronics industry. Sony, not Samsung was a household name in many parts of the world. Yet within the short span of less than seven years Korea announced in June 1987 the completion of a nationwide public switched telephone network (PSTN). The PSTN was more modern than the telephone networks in Japan and other advanced economies, which still relied heavily on older crossbar switching technology. Its completion was a major landmark in Korea's digital development that coincided with a dramatic move toward democracy and political liberalization just over a year before the highly successful 1988 Seoul Olympics.

Public-private partnership in industrial policymaking played a key role in Korea's ICT-driven development and continues to do so even today. However, as discussed in Chapter 3, the manner of such collaboration differed from Japan in several major ways.

Strategic restructuring

Wilson's strategic restructuring model also proves useful for our study both because it takes the diffusion of ICT as the dependent variable of interest and because it focuses on the experience of developing countries. His model seeks to explain ICT diffusion through the interaction of four distinct determinants:

- structures (especially social structures, but also economic and political structures);
- institutions (that is, persistent patterns of roles and incentives);
- politics (especially elite strategic behaviors); and
- government policies (specifically a mix of four policy balances – private and public initiative, competition and monopoly, foreign and domestic, centralized and decentralized).[40]

Wilson's model takes into account the role of leadership and vision in ICT diffusion. Also, the four policy balances provide a good framework for interpreting the changes in Korea over the past three decades. The strategic restructuring framework also acknowledges the important reality that shifts in these four balances within the ICT sector are highly political, with the politics determining which societal interests will get to control the richest and most politically sensitive sector in the modern world.[41]

Korea's geography and culture

As noted at the start of this chapter, Korea's mountainous geography exerted a strong influence on Korean culture over the millennia. Three other aspects of the nation's geography help to place its ICT-driven development in context.

First, Korea is a nation surrounded by islands. One source estimates that statistically, there are 3,421 islands surrounding the Korean Peninsula, 2,900 of which are found along South Korea's long coastline. Other estimates are higher, but this depends upon the definition of "island," given the many uninhabited small islets. The majority of South Korean islands are found along the western and southern coasts, and an estimated 705 of them are inhabited.[42] In addition to their geographical and historical relevance to Korea's culture, its islands bear a particular relationship to this book. The largest island, Jeju-Do has a land area of 1,825 square kilometers (705 square miles) and constitutes a semi-autonomous administrative district within Korea's government. Featuring an abundance of wind, it was chosen as the site of the nation's smart grid pilot project. The Korea Electric Power Corporation (KEPCO) initiated smart grid demonstration projects on approximately 30 of the other inhabited islands in South Korea and based on their success plans to dramatically expand the number of island-based smart microgrids. Chapter 7 will explore Korea's approach to green growth and its smart grid initiatives.

Second, Korea is situated as a peninsular nation bordered by larger neighbors, including Japan to the east and China to the north and west and a short border with Russia to the north. Over the course of history, China has exerted great influence over Korea, and Japan forcibly occupied the nation during the first half of the 20th century. At the end of World War II, Soviet forces entered Korea from the north, prompting U.S. to enter from the south and propose the 38th parallel as a dividing line.

Finally, the dominant feature of Korea's political geography today is the division of the nation, the world's most conspicuous unresolved legacy of the Cold War era. South Korea represents half of a divided nation, while North Korea occupies the other half. In the long sweep of history, Korea's division is obviously an aberration. Also, South Korea's digital development since the 1980s was not matched by North Korea, transforming the DMZ from simply a military demarcation line to the world's most poignant and deep digital divide.

Within Korea's largest urban concentrations, there is also abundant evidence for the manner in which culture along with geography influenced the arrangement of housing, markets and commercial services. The large high-rise apartment blocks in which most people live helped accelerate the construction of fiber optic broadband networks. Another example is Korea's tea-room (*tabang*) culture and long tradition of meeting friends outside the home. Today it is a booming coffee house culture, led by Starbucks, which as of 2013 operated 559 stores in South Korea, a number exceeded only in the U.S., Canada, Japan and China.[43] Joined by such franchises as Café Bene, Hollys Coffee, Ediya Coffee and Angel-in-US, the number of coffee shops in South Korea increased nearly ten-fold to 12,381 during the five years from 2006 to 2011.[44] Seoul has the highest density of coffee shops of any comparable metropolitan area in the world.[45] In a poignant contrast, back in the 1970s phone service in Korea was so expensive and inadequate that many small business owners resorted to using the phone number of a tea room on their business cards.

The historical context

Over the nearly four decades covered by this study, three secular trends influenced Korea's response to the global digital revolution. Globalization, urbanization and the rise of concern about environmental sustainability each helped shape the nation's strategic agenda and the restructuring of its ICT sector.

Globalization: export growth and global market pressures

Korea's approach to ICT strategy was strongly affected by mounting pressure from the United States, multinational companies and eventually the World Trade Organization (WTO) to liberalize its equipment and service markets. Market liberalization, a major theme of academic literature on telecommunications policy, implies privatization, government deregulation and the introduction of free market competition.

The Korean government had pursued export-led development beginning as early as the 1960s. If anything, the importance of exports increased under the so-called "developmental state" of President Park Chung Hee in the 1970s. Even today, South Korea stands out among the OECD economies for its heavy dependence upon the export of manufactured goods for economic growth.

The growth of Korean exports from the latter decades of the 20th century to the present inexorably increased the salience for Korea of international trade talks, beginning with the U.S. bilateral negotiations and proceeding through formation of the World Trade Organization. Accordingly, we will note the influence of international trade negotiations at certain key points on the strategic restructuring of South Korea's ICT sector.

Urbanization: the rapid rural to urban shift in Korea

Digital network development in South Korea was also powerfully shaped by a related secular trend, that of urbanization. The mountainous terrain contributed to the nation's dense pattern of urbanization, in which half the nation's population lives in and around the capital city of Seoul. The satellite photo in Figure 1.1 clearly shows the patterns of urban concentration in South Korea, particularly in the national capital area (*sudogwon* in Korean) around Seoul, which is now home to half the nation's population. As measured by its "urban footprint," defined as "the lighted area that can be observed from an airplane (or satellite) on a clear night,"[46] the national capital area around Seoul is the third largest urban area in the world.[47]

Urbanization in South Korea exhibited several other distinct characteristics. First, it occurred very rapidly,[48] leading some analysts to call it compressed urbanization. Indeed, until the mid-20th century South Korea retained many of the characteristics of a rural peasant society owing in part to its long colonial occupation by Japan.

Urbanization also led to a concentration of top corporate and educational institutions in Seoul and mirrored the development of all forms of modern transportation. As of 2012, Seoul had the third largest subway system in the world in terms of passengers served, behind Tokyo and Beijing.[49] The other urban concentrations shown in Figure 1.1 are mostly located along the nation's expressways and major railroad routes. Finally, as discussed by Choe and Kim, urbanization and globalization are interdependent. "As the globalization process deepens, this interdependency grows and becomes more complex, moving from simple trade interdependence to a complicated interdependence of global production, distribution, consumption and finance."[50]

Responding to both urbanization and the digital network revolution, in 2006 the Korean government released of the U-Korea master plan, the first such nationwide effort in the world to introduce sensors and ambient intelligence into an ever-denser digital network environment. In February of the same year the ministries in charge of ICT and construction-transportation signed an MOU aimed at building industry-wide partnerships between the ICT and construction sectors to integrate advanced IT infrastructure into the construction of sustainable cities. Around the same time, several city governments, including Seoul, Busan and Incheon, indicated their intent to pursue u-city development independently and all six regional governments planned u-city development projects. Consequently, in 2007 Korea passed a law on the construction of u-cities, which allowed the central government to exert some policy influence on the various local efforts.[51]

The emergence of sustainability: from brown growth to green growth policy in Korea

The growth of global concern with environmental sustainability was a third trend shaping Korea's digital development. The origin of this concern is often associated with the 1987 Brundtland Commission report, which defined sustainable development as "development that meets the needs of the present without compromising the ability of future generations to meet their own needs."[52] The internationally accepted framework that grew out of the Brundtland Commission report and discussions at the Earth Summits of 1992 and 2002 was made up of three elements, considered to be of equal importance. These were economic development, social development and environmental protection.[53] Some scholars have suggested the addition of two more dimensions, culture and governance, to this model. While these are important, we would suggest, as discussed in Chapter 7, that ICT as a critical enabler and infrastructure element for achieving sustainability, should be part of the model.

As a 2010 report by the International Institute for Sustainable Development put it succinctly, "Two issues of profound importance lie at the heart of current thinking about the development of global economies and societies: the challenge of environmental sustainability, and the potential of information and communications technology."[54] A later report by Souter notes these two important issues and then suggests that they together raise the central issue of "How far and in what ways do we need to change our understanding of sustainability in the light of the information and communications revolution?"[55] Indeed, it is difficult to even conceive of sustainable development without global communications and knowledge exchange. Nevertheless, there has been a surprising lack of interaction between policymakers and activists concerned with sustainable development, on the one hand, and those focused on ICT/internet public policy, on the other.[56]

The story of how Korea shifted from a brown growth policy in the early decades of its development, to a green growth strategy today, is a fascinating one. As of the 1980s Korea, with its manufacturing and export led economy, pursued a carbon intensive development trajectory. From 1990 to 2007 the nation's carbon emissions growth was the fastest in the OECD. Historically, Korea was also reluctant to sign international agreements on reducing carbon emissions, such as the Kyoto Protocol and Copenhagen Agreements. In spite of this history, in 2008 Korea made green growth a national policy priority.[57] The story of this shift in policy and Korea's overall approach to the challenge of sustainability will be told in Chapter 7.

The political, economic and social context for ICT-driven development

In addition to the secular trends discussed above, Korea's digital development occurred in the context of some major political, economic and social vicissitudes. A chronology of these changes begins with the Korean War armistice as a reminder that the nation's infrastructure and economy were nearly totally destroyed by the end of fighting in the Korean War. Under President Park Chung Hee the economy began to grow in the 1960s and 1970s, but by 1980 had not reached a takeoff point, as shown in Figure 1.3.

The 1980s began with Korea in a state of turmoil politically, economically and socially. During the final months of the Park Chung Hee government Korea's budding semiconductor industry was unable to compete with Japan and the U.S. in international markets, and color television broadcasting was not allowed. Also nothing had been done to solve the social problems caused by a massive backlog in provision of basic telephone service. President Park was assassinated in October of 1979.

By the summer of 1980 South Korea's economy had taken a nosedive. Korean politics took a leap forward toward democratization on June 29, 1987, the same month that the government announced completion of the nationwide PSTN. In the midst of political turmoil that threatened the successful hosting of the 1988 Seoul Olympics, Roh Tae Woo went on nationwide television and accepted virtually all demands of the opposition. This led to elections later that year in which Roh became president and to success in hosting the 1988 Seoul Olympics. Those games in turn were a powerful influence that helped Roh's "northern policy" of opening up trade, cultural and diplomatic ties with the Soviet Union, Eastern European nations, China and Vietnam, all of which had been completely cut off from South Korea during the long Cold War.

Almost exactly three decades after the 1988 Seoul Summer Olympic Games, South Korea hosted the 2018 Winter Olympics in Pyeongchang. While the Seoul Olympics took place at the peak of the 20th century mass media era, the Pyeongchang 2018 Winter Olympics occurred in the hyperconnected digital era. Beginning with the 2012 Olympics in London, the games are now referred to by some as the "bring your own device Olympics," transforming the Olympic experience for fans, athletes and the international media.

For Korea, both the 1988 and the 2018 Olympics, summer and winter, provide essential context for understanding its network-centric digital development. The modern Olympics, along with such events as World Cup soccer, are one of the world's leading media events, with the power to draw the attention of massive global audiences, especially for their opening and closing ceremonies. Television and the media play a central role in the Olympics, and consequently the Winter and Summer games provide regular opportunities for the host nation and corporate sponsors to display to the world the latest in network technology. In Pyeongchang Korea launched a pilot version of 5G mobile service, giving the world a glimpse of what to expect when international 5G standards are agreed upon in 2020.

As in Seoul three decades earlier, the Pyeongchang Winter Olympics played an important political and diplomatic role. North Korea accepted the invitation for some of its athletes to participate in the games, and there is evidence that the associated Olympic diplomacy played an important role in the warming relations between the two Koreas, leading to the summit meetings between President Moon Jae In and Chairman Kim Jong Un in 2018. Chapter 8 is devoted to the role of the Summer and Winter Olympics in Korea's digital development.

Two major international economic crises dramatically affected Korea's socio-economic development, and they show up in Figure 1.3 as the two visible dips in the overall trend toward a rapidly increasing GNI per capita. First came the Asian economic crisis of 1997–1998, widely referred to in Korea as the "IMF Crisis." Next was the global financial crisis of 2009. In both cases the nation was able to recover and resume its rapid growth.

In 2017 Korea experienced another political shock as President Park Geun Hye was impeached and removed from office. Moon Jae In won the May 10 presidential election and introduced some restructuring of ICT governance. For example, the former Ministry of Science, ICT and Future Planning was renamed the Ministry of Science and ICT.

An important social change that drove Korea's digital development was study abroad. As later chapters will discuss, Korea possessed a long cultural heritage of commitment to education and by 1980 had sent many of its best and brightest students abroad, mostly to the United States for graduate study. Quite a number of these individuals had returned to Korea and were highly placed throughout government, industry and academia. They included Ph.D.s in economics and various engineering fields who were well versed in the current state of technology in electronics, telecommunications and computing. Dr. Oh Myung mobilized a number of these individuals and led what is now referred to as the "triumph of the technocrats" beginning in the 1980s.

A major tragedy in South Korea also helped shape its digital development. In April of 2014 a ferry traveling from Incheon to Jeju Island and carrying hundreds of high school students on a field trip capsized and sank, killing most of the students. The incident was widely covered by the media, but more importantly all of the students were carrying LTE phones, so the last-minute phone calls to friends and immediate family, text messages and recorded videos brought home the heart-wrenching nature of this tragedy. The sinking of the Sewol Ferry seemed to transfix the nation for months as it became the focus of political attention and even caused an economic slowdown.[58]

In the immediate aftermath of the Sewol tragedy, the need for a unified public safety network was brought up in the national assembly, and the government initiated a project to construct such a network a month later. In July of 2014 the Ministry of Science, ICT and Future Planning selected Public Safety LTE (PS-LTE) as the technology for the new network, and in November of the same year the Ministry of Public Safety and Security started implementation with the aim of constructing the nationwide network by the end of 2018.[59] The SafeNet Forum was also created in 2014 as an official body to reflect the opinions of various parties and to monitor the government's policies in order to maintain fairness and transparency in the process of completing construction of a nationwide public safety network.[60] The forum includes all of the major players from industry, academic and government. Chapter 9 will explore Korea's progress in building one of the world's first nationwide public safety networks and its active role in international efforts to develop standards for such new networks.

The "Miracle on the Han" in context: a network-centric perspective

This book presents a network-centric perspective on Korea's remarkable digital development from 1980 to the present. Such an approach should help an international audience better understand the nation's ICT-driven development it in several ways. First, it helps to situate South Korea's ICT-driven development for an international audience in terms of today's terminology, most especially the recognition of the ICT sector as an engine of economic growth and a necessary part of sustainable development.

Second, it clearly identifies the origins of digital development in Korea's "1980s telecommunications revolution," coinciding with the global introduction of digital switching. That historical coincidence is vital to an understanding of the power and longevity of this transformation, given the long-term, cumulative nature of broadband network infrastructure projects. Ironically, Korea's digital development originated in the midst of crises affecting both the nation's telephone network and its broadcasting networks. These are a major subject of Chapter 2. As this book will chronicle, by building digital networks in the 1980s, then taking the lead in fixed and mobile broadband networks in the 1990s Korea literally became a network society years in advance of most other nations.

Third, a network-centric approach conveys that the main locus of innovation in South Korea's ICT sector since the 1980s was in its ever-advancing digital networks. This sheds light on the role of digital network infrastructure investment and on hardware manufacturing versus software and services in building the network society.

Fourth, our approach helps to explain the important relationship of Korea's human networks to its digital ones. Korean culture itself seems conducive to the development and use of digital networks as it is highly attuned to the interpersonal networks that formed the basis for effective public-private partnership and leadership, including what Jho called "centralized network governance" as a steering mechanism for policy.[61] Also, the network of overseas Koreans, estimated in 2017 to number nearly 7.5 million,[62] played an important role in the 1970s efforts to develop the electronics industry and in later developments in semiconductors and digital switching in the 1980s.

Fifth, a network perspective helps to clarify the social and political impact of the digital disruption in Korea. The book will document examples ranging from the invention of StarCraft, the world's first massive multiplayer online game (MMOG), to the introduction of social media in the form of Cyworld, to the more recent tragic sinking of the Sewol Ferry in 2014 which galvanized support for building what will likely be the world's first dedicated, nationwide public safety LTE (PS-LTE) network.

Sixth, this study underscores the importance of long-term planning, investment and government leadership in building digital network infrastructure. In retrospect the comprehensive *Long-Term Plan to Foster the Electronics Sector* drafted in 1980 and 1981 ensured that the nation would develop strength in manufacturing most of the key components that make up the nuts and bolts structure of the internet and related digital networks. These included switches, routers and a growing array of end systems that include television sets, computers, tablets, smartphones and IoT devices.

Finally, a network-centric perspective proves valuable when contemplating the prospects for future networks. Current policy discussions about digital technologies center on such topics as robotics, augmented reality (AR), virtual reality (VR), the Internet of Things (IoT), artificial intelligence and blockchain. All of these technologies rely heavily on the increasing capabilities of digital networks. Taking them all into account, it appears that the network itself will continue to be an important locus of innovation. The concluding chapter of this book will speculate about the role of future networks, in Korea and globally.

Notes

1 Choe, Chungho. (1994). Korea's landscape and mindscape. *Koreana*, 8(4) (Winter). Retrieved from www.koreana.or.kr
2 Kelly, Tim, Vanessa Gray and Michael Minges. (2003, March). *Broadband Korea: Internet case study*. International Telecommunications Union, p. 1.
3 *Republic of Korea transition to a knowledge-based economy*, Report No. 20346-KO. World Bank East Asia and Pacific Region, June 29, 2000.
4 International Telecommunications Union. (2015). *Measuring the information society report 2015*. Geneva: ITU. Retrieved May 15, 2016, from www.itu.int/en/ITU-D/Statistics/Documents/publications/misr2015/MISR2015-w5.pdf
5 International Telecommunications Union. (2015). *Measuring the information society report 2015*. Geneva: ITU. Retrieved May 15, 2016, from www.itu.int/en/ITU-D/Statistics/Documents/publications/misr2015/MISR2015-w5.pdf
6 Marx, Axel and Jadir Soares. (2013). *South Korea's transition from recipient to DAC donor: Assessing Korea's development cooperation policy*. International Development Policy, pp. 107–142. Retrieved June 28, 2016, from https://poldev.revues.org/1535
7 NASA Visible Earth astronaut photo taken January 30, 2014. Retrieved November 5, 2018, from https://visibleearth.nasa.gov/view.php?id=83182
8 Isaacs, Harold R. (1958). *Scratches on our minds: American images of China and India*. New York: The John Day Company, p. 37. Full text Retrieved November 5, 2018, from https://archive.org/stream/scratchesonourmi010380mbp/scratchesonourmi010380mbp_djvu.txt
9 Effective Language Learning. (2016, October 30). Language difficulty ranking. Retrieved from Effective language learning: www.effectivelanguagelearning.com/language-guide/language-difficulty

10 Kelly, Tim, Vanessa Gray and Michael Minges. (2003, March). *Broadband Korea: Internet case study*. International Telecommunications Union, p. 2.

11 Oh, M. and J. F. Larson. (2011). *Digital development in Korea: Building an information society*. London: Routledge, p. 19.

12 Kelly, Tim, Vanessa Gray and Michael Minges. (2003, March). *Broadband Korea: Internet case study*. International Telecommunications Union, p. 6.

13 Oh, M. and J. F. Larson. (2011). *Digital development in Korea: Building an information society*. London: Routledge, p. 29.

14 Fransman, M. (2010). *The new ICT ecosystem: Implications for policy and regulation*. Cambridge: Cambridge University Press, pp. 10–11.

15 Fransman, M. (2010). *The new ICT ecosystem: Implications for policy and regulation*. Cambridge: Cambridge University Press, p. 89.

16 OECD. (2016). *OECD broadband portal*. Retrieved May 15, 2016, from OECD: www.oecd.org/sti/broadband/oecdbroadbandportal.htm

17 This chapter draws heavily on Larson, James F. (2017). Network-centric digital development in Korea: Origins, growth and prospects. *Telecommunications Policy*, 41, pp. 916–930.

18 Haviland, David B. (2002). The transistor in a century of electronics. Retrieved April 17, 2016, from Nobelprize.org: www.nobelprize.org/educational/physics/transistor/history/

19 Shannon, Claude. (1948). A mathematical theory of communication. *Bell System Technical Journal*, pp. 379–423.

20 Kurose, J. F. and Keith W. Ross. (2013). *Computer networking: A top-down approach* (6th ed.). Essex: Pearson Education Limited, pp. 27–31.

21 Benkler, Y. (2006). *The wealth of networks: How social production transforms markets and freedom*. New Haven, CT: Yale University Press, p. 32.

22 Benkler, Y. (2006). *The wealth of networks: How social production transforms markets and freedom*. New Haven, CT: Yale University Press, pp. 36–37.

23 Hilbert, M. A. (2011). The world's technological capacity to store, communicate and compute information. *Science*, pp. 60–65. doi:10.1126/science.1200970

24 Norris, Pippa. (2001). *Digital divide: Civic engagement, information poverty and the internet worldwide*. Cambridge: Cambridge University Press.

25 Mansell, R. (2008). The life and times of the information society: A critical review. *Media@LSE fifth anniversary conference: Media, communication and humanity 2008*. London: LSE department of media and communications, p. 26.

26 Pool, Ithiel de Sola. (1983). *Technologies of freedom*. Cambridge, MA: Harvard University Press, p. 6.

27 Pool, I. D. (1974). The rise of communications policy research. *Journal of Communication* (Spring), pp. 31–42.

28 Pool, I. D. (1974). The rise of communications policy research. *Journal of Communication* (Spring), p. 33.

29 Pool, I. D. (1974). The rise of communications policy research. *Journal of Communication* (Spring), p. 39.

30 Bell, D. (1973). *The coming of post-industrial society*. New York: Basic Books.

31 Bell, D. (1973). *The coming of post-industrial society*. New York: Basic Books, p. xxxvii.

32 Castells, M. (2010). *The rise of the network society* (2nd ed.). West Sussex: Wiley-Blackwell.

33 McLuhan, M. (2003). *Understanding media: The extensions of man*. Berkeley: Gingko Press, Critical Edition.

34 Jho, W. (2007). Liberalization as a development strategy: Network governance in the Korean mobile telecom market. *Governance: An International Journal of Policy, Administration and Institutions*, 20(4).

35 Jho, W. (2007). Liberalization as a development strategy: Network governance in the Korean mobile telecom market. *Governance: An International Journal of Policy, Administration and Institutions*, 20(4), p. 636.

36 Rogers, E. M. (2003). *Diffusion of innovations* (5th ed.). New York: Free Press, p. 12.

37 Rogers, E. M. (2003). *Diffusion of innovations* (5th ed.). New York: Free Press, p. xviii.

38 Johnson, C. (1982). *MITI and the Japanese miracle: The growth of industrial policy, 1925–1975*. Stanford: Stanford University Press.

39 Johnson, C. (1999). The developmental state: Odyssey of a concept. In Meredith Woo-Cumings, *The developmental state*. Cornell: Cornell University Press, pp. 37–39.

40 Wilson, E. J. (2004). *The information revolution and developing countries*. Cambridge, MA: The MIT Press, p. 39.

41 Wilson, E. J. (2004). *The information revolution and developing countries*. Cambridge, MA: The MIT Press, p. 104.

42 Soh, Chin Thack. (1980). *Korea: A geomedical monograph of the Republic of Korea*. Berlin Heidelberg: Springer-Verlag, p. 6.

43 Starbucks Corporation. (2014). Starbucks corporation fiscal 2013 annual report. Starbucks Corporation.

44 Jin, H. (2012, August 21). Korean coffee craze may be hit by curbs. *Reuters*. Retrieved April 8, 2014, from http://uk.reuters.com/article/2012/08/21/us-korea-coffee-idUKBR E87K02W20120821

45 Wanninger, J. (2013). *A coffee love story*. Retrieved April 7, 2014, from Invest Korea: Korea's National Investment Promotion Agency: http://blog.investkorea.org/wordpress/?p=4134

46 Demographia. (2013). *Demographia world urban areas (world agglomerations)* (9th annual ed.). Belleville, IL: Demographia, p. 2. Retrieved November 16, 2013, from www.demographia.com

47 Demographia. (2013). *Demographia world urban areas (world agglomerations)* (9th annual ed.). Belleville, IL: Demographia, p. 16. Retrieved November 16, 2013, from www.demographia.com

48 Kang, Myung Goo. (1998, June). Understanding urban problems in Korea: Continuity and change. *Development and Society*, 27(1), pp. 100–101.

49 Bae, H. J. (2013, November 15). Subway mirrors urban, national growth. *The Korea Herald*. Retrieved November 17, 2013, from www.koreaherald.com/view. php?ud=20131115000831

50 Choe, Sang Chuel and Won Bae Kim. (2001). Globalization and urbanization in the Republic of Korea. In S. S. Yusuf, *Facets of globalization: International and local dimensions of development World Bank discussion paper No. 415*. Washington, DC: World Bank.

51 Oh, M. and J. F. Larson. (2011). *Digital development in Korea: Building an information society*. London: Routledge, pp. 113–126.

52 United Nations. (1987). *Report of the World Commission on Environment and Development: Our common future*. United Nations. Retrieved from www.un-documents.net/our-common-future.pdf

53 Retrieved from December 11, 2018, from https://en.wikipedia.org/wiki/File:Sustainable_development.svg#file

54 Souter, David, Don MacLean, Ben Akoh and Heather Creech. (2010). *ICTs, the internet and sustainable development: Towards a new paradigm*. Winnipeg: International Institute for Sustainable Development (IISD). Retrieved September 30, 2015, from www.iisd.org/pdf/2010/icts_internet_sd_new_paradigm.pdf

55 Souter, D. (2012, May). *ICTs, the Internet and sustainability: A discussion paper*. Winnipeg: International Institute for Sustainable Development.

56 Souter, David, Don MacLean, Ben Akoh and Heather Creech. (2010). *ICTs, the internet and sustainable development: Towards a new paradigm*. Winnipeg: International Institute for Sustainable Development (IISD), p. 5. Retrieved September 30, 2015, from www.iisd.org/pdf/2010/icts_internet_sd_new_paradigm.pdf

57 Kim, S. Y. (2015). Developmental environmentalism: Explaining South Korea's ambitious pursuit of green growth. *Politics & Society*, pp. 1–28. doi:10.1177/0032329 215571287

58 Lee, J. (2015, April 8). Korea economic sentiment yet to fully recover from ferry sinking. Retrieved June 4, 2016, from *Bloomberg Markets*: www.bloomberg.com/news/articles/2015-04-07/korea-economic-sentiment-yet-to-fully-recover-from-ferry-sinking

59 Hong, Daehyoung. (2016, 5). Greetings from SafeNet Forum Chair. Retrieved June 4, 2016, from SafeNet Forum: http://safenetforum.or.kr/eng/main/main.php?categoryid=02&menuid=01&groupid=00#

60 SafeNet Forum. (2016, 6). *Background and objectives*. Retrieved June 4, 2016, from SafeNet Forum: http://safenetforum.or.kr/eng/main/main.php?categoryid=02&menuid=02&groupid=00

61 Jho, W. (2007). Liberalization as a development strategy: Network governance in the Korean mobile telecom market. *Governance: An International Journal of Policy, Administration and Institutions*, 20(4), p. 636.

62 Ministry of Foreign Affairs and Trade, 2017 재외동포현황(180125).

2 The origins of network-centric digital development in Korea

"Truly, if we want to invigorate the electronics industry you must go to the Ministry of Communications. The core of the electronics industry is telecommunications."

Dr. Kim Jae Ik to Dr. Oh Myung, 1981

An in-depth understanding of digital development in Korea begins with an understanding of when and how it originated. As noted in the opening chapter, Korea was a striking exception to the general pattern in which adoption of digital technologies occurred at a faster pace in advanced, industrialized economies than in the developing world. Korea's early start on the construction of nationwide digital networks is a crucial point for at least two reasons. First, building the digital "information superhighways" is an expensive, long-term project. Second, innovations in networking technology are cumulative, subject to the "on the shoulders of giants" characteristic of information itself.

For South Korea, the origins of the digital transformation can mostly be traced to a set of circumstances and policies adopted and implemented during the politically turbulent 1980s. During that remarkable decade key institutional and political changes took place that effectively jump-started the nation's digital development, allowing it to harness the power of the network revolution in its early stages. It started with a massive backlog in the provision of basic, non-digital phone service and ended with same day installation service and one of the most modern digital telecommunications networks in the world. These events and others led knowledgeable Koreans to refer to "the 1980s telecommunications revolution."

The purpose of this chapter is to explain the origins of Korea's digital development based on historical evidence. It addresses the questions of why, how and when South Korea set out on a path of network-centric and ICT-led development and places these developments in a global context. The analysis is based on insider accounts and Korean language sources as well as English language research.

Precursors of digital takeoff

As noted in the opening chapter, three 20th century breakthroughs in technology and theory laid the groundwork for the global digital network revolution. First came the invention of the transistor, which was followed by Shannon's mathematical theory of communication, eventually leading to the development of fully electronic switches. In addition to these important developments outside Korea, there were other important precursors of ICT-driven development that took place inside Korea.

Policies under Park Chung Hee

In several important ways, the government of President Park Chung Hee laid the foundations for the nation's ICT-driven economic takeoff in the 1980s. Policies and plans were implemented and key institutions created that would play a central role in later development. Equally as important, lines of decision-making within government and the pattern of public-private partnership of industry with government were firmly established.

Korea's economic development in the 1960s and 1970s was government led and export-oriented. The average annual growth rate over these two decades was almost 10 percent. In the 1970s the nation's industrial policy emphasis shifted from light industry to the heavy and chemical industries (HCI), which included electronics. The 1970s featured oil shocks in 1972 and 1979, which hit Korea hard, given its energy intensive industrial strategy. In a decade of global inflation, Korea's annual inflation rate stood as one of the highest in the world at 21 percent.[1]

Korea's GNI per capita began to rise in the latter part of the 1970s in no small part because of several initiatives under the Park Chung Hee administration that laid some of the groundwork for later digital development and rapid economic growth. For example, the Electronics Industry Promotion Law No. 2098 was promulgated in 1969, as an outgrowth of the heavy and chemical industry and defense initiatives of the Park administration. The last five-year plan outlined by his administration followed an earlier pattern of industrial policy in Japan and shifted emphasis toward development of the nation's heavy and chemical industries, including electronics.

Another measure under the Park Chung Hee administration was the establishment of research and teaching institutions to lead the development of scientific and engineering fields in Korea. The first of these was the Korea Institute of Science and Technology (KIST), agreed upon in a joint public statement by Korean president Park Chung Hee and U.S. president Lyndon Johnson in 1965 and implemented the following year through an agreement between Korea's Economic Planning Board and the U.S. Agency for International Development (U.S. AID). Its mission was to play a central role as the first comprehensive research agency to promote the nation's economic growth and the modernization of engineering fields.

In terms of organization and management, KIST was a groundbreaking effort for Korea, modeled after some of the world's most progressive research institutions, including the National Research Council of Canada, the Commonwealth Scientific and Industrial Research Organization of Australia and the Max-Planck Institute of Germany.[2]

Although KIST marked a major step forward for Korea in science and technology it did not directly address the need to increase the supply of researchers since it did no teaching. Consequently, another major institution, the Korea Advanced Institute of Science (KAIS), which later became the Korea Advanced Institute of Science and Technology (KAIST), was established in 1971.

As with KIST, the initial planning and funding for KAIS was conducted in cooperation with the U.S. AID. A survey team formed by that agency was chaired by Frederick Terman, the former dean of the school of engineering and Provost at Stanford University who is widely regarded as the father of Silicon Valley. The Survey Report on the Establishment of the Korea Advanced Institute of Science, informally referred to as the Terman Report, noted that

> At the graduate level, education in science and engineering in the U.S. sense is almost entirely lacking in Korea, and what there is is highly fragmented. Thus in 1969 the nation's 600 graduate students studying for the MS degree in science and engineering were distributed among 152 departments in 22 different schools! Korea graduate schools in science and engineering appear to have had little if any impact on the Korean economy. Many Korean students go abroad for graduate work, but few return. However, even if they did return, their foreign training would be oriented toward the needs of a developed nation, not those of Korea. Korea now lacks the technological base required to provide independent indigenous strength to Korean industry. A self-sustaining Korean economy needs a steady supply of engineers and applied scientists who combine high ability with advanced training oriented toward the technological needs of Korean industry.[3]

In order to meet the need addressed by the Terman Report, the basic law and planning documents for KAIS included a number of provisions that were unprecedented for Korean institutions of higher education at that time. These included provisions ensuring that KAIS:

- will not come under existing rigid educational laws and public employees acts;
- will have stable support through income from an endowment provided by the government;
- will be empowered to recruit and support faculty on terms that will make it possible to bring back to Korea well qualified scientists and engineers now abroad;
- will ensure students receive generous financial support, be provided with dormitory facilities and will receive special treatment with respect to military service;
- will have an independent self-perpetuating board of trustees that will exercise full responsibility for it; and
- is authorized to confer Doctor of Science (ScD), Engineering and Master of Science (MS) degrees in accordance with its own regulations.[4]

Another important antecedent of strong and sustained digital development in Korea was the manner in which the government recruited and relied upon technocrats, rather than career civil servants in the process of formulating and implementing policy for the electronics industry. A leading example was O Won Chol who served as the senior presidential secretary for economic affairs to President Park Chung Hee for eight years in the 1970s. He was trained as a chemical engineer and, partly for that reason, has been referred to as "Korea's first technocrat." O spearheaded Korea's shift from light industry to heavy and chemical industries in the mid-1970s, following the example of Japan's earlier shift in industrial policy aimed at increasing exports.[5]

President Park Chung Hee also enlisted the advice of Columbia University professor of electrical engineering Kim Wan Hee. Professor Kim, who served as a professor of electrical engineering at Columbia for 21 years, served as an advisor to President Park on matters pertaining to the electronics sector. The direct and personal nature of this advice is indicated by both the number of trips he made to Korea and by the remarkable personal correspondence sent to him by President Park. In fact, Professor Park donated the 122 handwritten letters from President Park Chung Hee to Korea's national archives. The subject of this correspondence was the "electronics industry." In one letter sent in 1968, Park wrote, "I wish to hear from you and hope for blessings for your family. Respectfully, Park Chung Hee." In those letters, he didn't use his presidential title and wrote with the utmost respect.[6]

Also during the 1970s, a group of technocrats who would be most influential in the post-Park era joined the Korean government. One of these, Dr. Kim Jae Ik, majored in international relations at Seoul National University, pursued a master's degree in economics at the University of Hawaii's East-West Center and completed a Ph.D. in economics at Stanford in 1973. During his years at Stanford he renewed acquaintance with Professor Nam Duck-Woo of Sogang University who was on sabbatical at Stanford and with whom Dr. Kim had worked at the Bank of Korea after graduating from SNU.[7]

In 1974 Kim Jae Ik made a study trip to Taiwan,

> where he found the electronic digital telephone switch system to be far more efficient than Korea's mechanical switching system. He felt that in the long run Korea would benefit greatly from an electronic switching system and strongly recommended its introduction. In doing so, he encountered serious opposition to a policy change for the first time in his government career. In opposing the shift to an electronic system, the local manufacturers of telecommunications equipment went so far as to question Kim Jae Ik's personal integrity. Nevertheless, the change was made in 1980, which not only improved telephone service throughout the country but also provided Korea with an updated communications infrastructure and technology that has been vital in maintaining its competitiveness in the international markets.[8]
>
> Kim Jae-Ik also had a remarkable combination of personal characteristics that made him effective in a governmental setting; he was an easy person to

agree with. He was genuinely modest and soft-spoken, but that did not keep others from recognizing that he had a brilliant mind. He could persuasively and persistently argue for his positions. He knew how to marshal facts, figures, and logic for his case for the purpose of convincing others rather than overwhelming them. . . .

Finally, Kim Jae-Ik had deep philosophical convictions and a vision of the future of his country. He never doubted the superiority of the market over government fiat. To him, the market was not only more efficient in resource allocation but in the long run most conducive to the advancement of democracy. Furthermore, he believed that with liberal economic reforms and efficient management, Korea could one day emerge as a world-class industrial power. These beliefs and visions gave him tremendous strength and confidence. At all times he was serene, and to the end he was an optimist.[9]

This description of Kim Jae Ik, based on published accounts, helps set the stage for the Dr. Oh Myung's meeting with him in 1980. That dinner meeting, described later in this chapter, would have profound ramifications for the future of Korea's ICT sector.

In retrospect, it is clear that the Park Chung Hee administration shaped the essential character of public-private partnership in Korea's industrial policy. However, some of President Park's policies, notably his opposition to color television broadcasting, and the longstanding practice of appointing only government bureaucrats to ministerial positions, actually ran counter to the development of the electronics industry.

The world will never know what might have happened if Park had continued in office. His assassination in October of 1979 brought dramatic changes in the politics, economics and policy possibilities for Korea in the 1980s.

Politics, economics and society in 1980

As the decade dawned, Korea was a nation in crisis. Politically, a new government took control in the wake of President Park Chung Hee's assassination. Economically, an unusually cold and damp summer led to a disastrous harvest that year and agricultural production dropped by no less than 22 percent. The Korean economy suffered negative growth for the first time in two decades, contracting by 5.2 percent, while inflation soared to 29 percent in consumer prices and 39 percent in wholesale prices. The nation's current account deficit ballooned to over US$5.3 billion.[10] The sharp dip in economic growth is vividly illustrated in Figure 2.1.[11] The only other time the nation's annual GDP growth rate dipped into negative territory was during the Asian economic crisis of 1997–1998, known in Korea as the "IMF Crisis." Even during the global financial crisis of 2009 it remained positive.

Less well known around the world is that the nation's crisis in 1980 extended to its basic communication networks, both for telephone service and for broadcasting. In the telecommunications sector, all policymaking, regulation and provision of

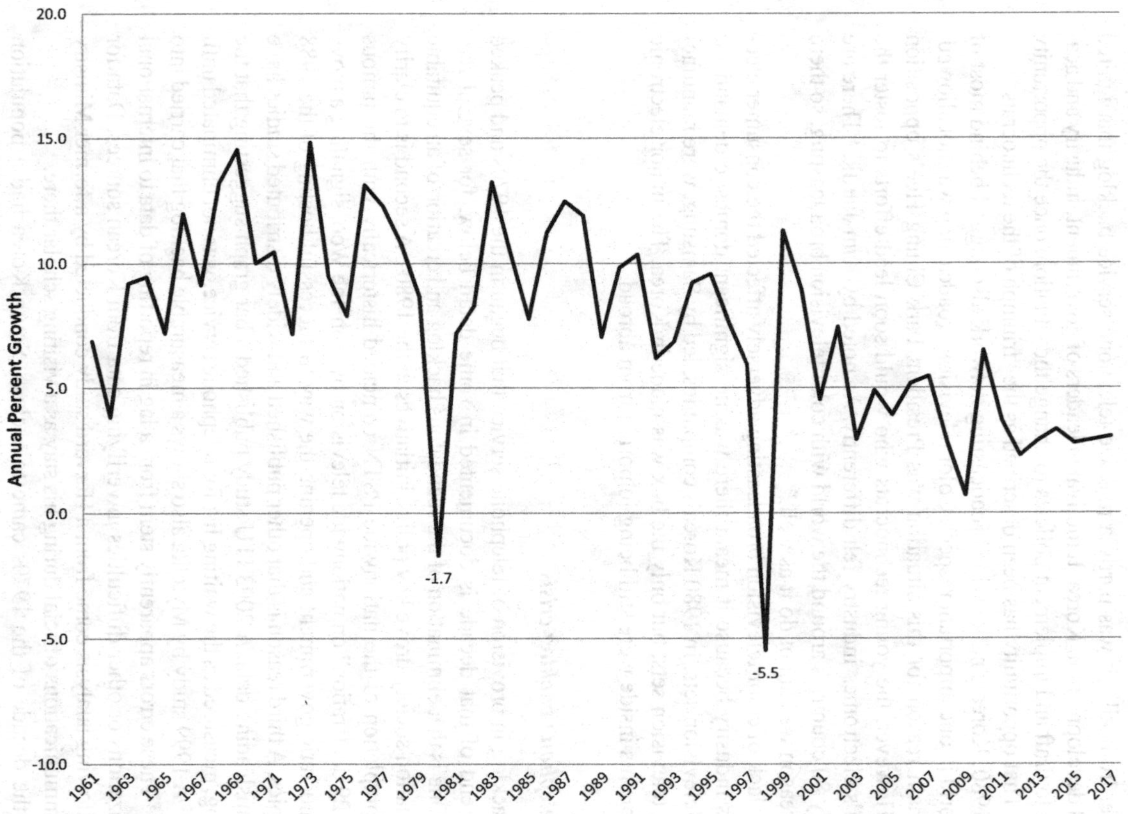

Figure 2.1 Korea's GDP growth (annual %)

Source: World Bank.

services were handled by the government monopoly Ministry of Communications (MOC). The MOC was a relatively weak Ministry whose Minister had traditionally been a political appointee. Korea at that time had no private telecommunications business, and even telephone handsets had to be purchased through the MOC.

As of 1980, the MOC found itself unable to meet the demand for basic telephone service. Ironically, it was partly a massive telephone service backlog that ignited digital development in Korea. It motivated leaders of government, industry and academia to draft and implement policies to change the situation, once the opportunity arose. That opportunity has been described as the "triumph of the technocrats."

In 1980, Korea's nationwide broadcasting network also lagged behind most of the world in one important respect. Color television broadcasting was not allowed. The main reason for this situation was President Park Chung Hee's opposition to it. However, the young technocrats who would soon lead efforts to foster the nation's electronics industry felt differently. As noted by Kim Jae Ik, "There are already 80 countries around the world with color television broadcasting, so there is no reason we cannot do it as well."

The lack of color television broadcasting adversely affected the consumer electronics industry because it meant there was no significant domestic demand for color television sets. In 1980 Korea's companies, led by Samsung, were manufacturing television sets, but only the box was made in Korea. The major electronic components inside were still being imported from abroad.

The telephone backlog crisis

The backlog in provision of telephone service that began in the 1970s and peaked at the end of that decade is documented in some detail below, for several reasons. One is that elimination of the scandalous backlog and creation of an equitable information society drove key policy initiatives in the 1980s. A second is to clarify that completion of the nationwide PSTN coincided historically with the famous "June 29 declaration" on nationwide television by Roh Tae Woo, signaling a move to democratic government and opening the way to successful hosting of the 1988 Olympics. A third reason is that other published and widely distributed studies have used misleading data. A 2003 ITU study published a bar graph suggesting that the backlog, measured as the waiting list for telephone service, was not eliminated until 1989.[12] A 1999 study by Mytelka also shows a measurable backlog that carried into 1988.[13] These errors apparently stem from a lag in reporting of data to international organizations or other difficulties in verifying data from Korean sources. Data for the following analysis come from a Harvard study conducted by a former Ministry of Communications official during his stay as a visiting scholar there.

As the decade of the 1970s came to a close, South Korea had a population approaching 36 million people, and its industrial and commercial base was rapidly developing, yet it had fewer than 2.8 million telephone lines. More than 600,000 of these lines, nearly one quarter of the installed capacity, remained on back order. This shortfall in telephone service created enormous social, economic and political problems. Figure 2.2 documents the evolution and scope of the problem.[14] Ironically, it was the success of Korea's third National Five-Year Economic

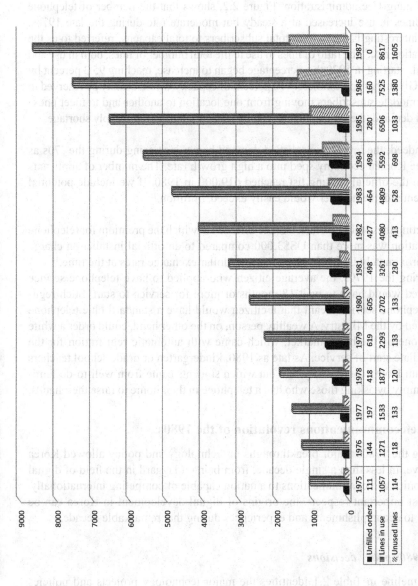

	1975	1976	1977	1978	1979	1980	1981	1982	1983	1984	1985	1986	1987
■ Unfilled orders	111	144	197	418	619	605	498	427	464	498	280	160	0
▥ Lines in use	1057	1271	1533	1877	2293	2702	3261	4080	4809	5592	6506	7525	8617
▨ Unused lines	114	118	133	120	133	133	230	413	528	698	1033	1380	1605

Figure 2.2 The telephone backlog crisis, 1975–1987 (units = thousands of lines)

Source: Based on Sung, K. J. (1990). *Korean telecommunications policies into the 1990s.* Cambridge, MA: Harvard University Program on Information Resources Policy.

Development Plan (1972–1976) that helped increase public demand for telecommunications services, thereby increasing the backlog problem.

As briefly noted in Chapter 1, systematic investment in telecommunications only began with the series of five-year economic development plans introduced by the Park Chung Hee administration. Figure 2.2, shows that the number of telephone main lines in use increased at a steady but moderate rate during the late 1970s. More interesting is the ratio of total subscribers to total capacity, referred to as the "utilization rate," the ratio of lines in use to the total number of lines, both in use and unused. Starting in 1970 this percentage began to increase, reaching 95.3 percent by 1980. Given that approximately 10 percent of capacity would usually be reserved to accommodate subscribers moving from one location to another and to meet unexpected demand, the utilization rate can be used as a measure of supply shortage.

> Indeed, the level of unsatisfied demand began increasing during the '70s as the Korean economy sped into a high growth rate. The number of applicants on the official waiting list reached 619,000 in 1980. If we include potential demand, the number would easily exceed 1 million.[15]

Reflecting these facts, a black market emerged in which the premium for telephone installation was more than US$3,000 compared to an official installation charge of approximately US$700, given the won/dollar exchange rates at the time.[16]

During the 1970s, the average citizen who applied to have telephone service installed would have to wait 18 months or more for service to start. Such regular telephone service meant that a citizen would have a standard blue telephone installed by the Ministry. A wealthy person, on the other hand, could order a white telephone on the black market, which came with automatic registration for the immediate start of service. As late as 1980, kindergarten or grade school teachers in South Korea could easily find out which students came from well-to-do families simply by asking those who had a telephone in their home to raise their hands.

The telecommunications revolution of the 1980s

During the 1980s major breakthroughs in technology and policy allowed Korea to move, in less than a single decade, from being a laggard in the field of digital electronics and communications to a nation capable of competing internationally. In most important respects, the origins of digital development in Korea can be traced to accomplishments and experiences during that remarkable decade.

Key 1980s policy decisions

The timeline in Table 2.1 identifies the major technology projects and policies that enabled Korea to accomplish so much in a single decade. Four key policy decisions in the 1980s deserve special emphasis. Korea's leadership decided to:

- develop and manufacture electronic switching systems (TDX);
- enter the global semiconductor industry, with an emphasis on DRAM chips;

Table 2.1 Policy and technology innovation milestones of the 1980s

1981	Completion of *Long-Term Plan to Foster the Electronics Sector*; color TV broadcasting introduction
1982	January – Formation of Korea Telecommunications Authority (KTA) and DACOM; TDX electronic switching project launch
1983	March – Deregulation of public switched telephone network (PSTN)
1984	Completion of digital toll switching network; Korea Mobile Telecommunications Corporation (KMT) founded
1985	September – Deregulation of the supply of terminals
1986	4MB DRAM semiconductor project
1987	June – Completion of PSTN
1988	September-October – Successful hosting of Seoul Olympics

- begin color television broadcasting; and
- separate the telecommunications business from the government ministry, including the formation of KTA (the Korea Telecommunications Authority, predecessor of KT (Korea Telecom)) in 1981, DACOM Corporation for data communications in 1982 and Korea Mobile Telecommunications Corporation in 1984.

From today's perspective, nearly four decades later, these decisions seem remarkably farsighted. They also explain why we characterize Korea's ICT-led development as "network-centric." Each decision anchored one part of South Korea's strong export-led economic development. The nation became a major manufacturer and exporter in each of the four industries affected by the policy decisions: advanced networks, semiconductor memory chips, flat screen color television sets and displays and mobile handsets.

These four industries and their underlying technologies undergird the hyperconnected network society in several crucial ways. First, they are technically closely interrelated and synergistic. Switches are essentially the computers that make today's digital networks possible, including the internet and the Internet of Things, the mobile apps ecosystem, cloud computing and big data. Semiconductors are, of course, an essential component in computers and a host of other electronic devices, including switches, television sets and mobile phones. Second, Moore's Law, with smaller, cheaper and more powerful semiconductors, fueled the mobility revolution, a pervasive development with profound consequences for how all of us communicate, today and tomorrow. Finally, advances in color displays and the emergence of the multi-screen world appeal to the all-important human sense of sight and promise within this century to extend other human senses through augmented and virtual reality that draws on the power of the new networks.

The triumph of the technocrats

Kim Jae Ik had held a senior working-level position at the Economic Planning Board (EPB) when he first met General Chun Doo Hwan in January of 1980. During his three years at the EPB his office was a primary stop for international visitors from World Bank officials to businessmen of every caliber. Foreigners liked Kim Jae Ik because of his clear-headed, no-nonsense approach to the Korean economy. Kim was fond of analyzing issues simply in terms of supply and demand, a concept that did not fit well with the visionary projects of Park Chung Hee and his nation-building enterprises of the 1960s and 1970s.[17] The January 1980 meeting between the director general of the Economic Planning Board's planning bureau and the two-star head of the Defense Security Command was "the start of an extraordinary alliance between a technocrat and a general, one that was to have a powerful impact on the direction of the Korean economy in the 1980s."[18]

Later, when Kim Jae Ik accepted the offer to become the senior secretary to the president, he became the top economic policy decision-maker in the Chun government. According to one account, right after Chun assumed the presidency in September 1980 he

> wanted to appoint Kim as the Senior Secretary. At the time, Kim asked Chun that he would accept the appointment under one condition, "if you are to implement economic policies as I advise you, you will have to face grave oppositions from almost everyone out there. Would you be willing to accept and implement my words against all others?" Chun's reply was simple and to the point: "No need to say anything else. You are the President when it comes to economic policies."[19]

According to the same account,

> The bottomless trust between President Chun and Secretary Kim developed through an unusual personal relationship they had with each other. The very first contact between Chun and Kim . . . took place when Chun was searching for his private tutors on economic policy areas during his chairmanship at the National Security Council (NSC). Then, Kim was about to relocate himself from the Director of Economic Planning Department of the EPB to KDI due to his idealistic attitudes towards economic policies and subsequent barriers that he felt during his tenure at the government. After Chun called on Kim, Kim spent at least two hours almost every day commuting to Chun's residence to tutor him about the basic economic principles towards current economic issues and policy solutions. This relationship continued after Chun entered the Blue House (the residence of the president), sometimes spending the whole day teaching Chun about economic policies.[20]

A momentous dinner meeting

During August of 1980 leaders from the Military Academy, members of the Cabinet and government department heads were summoned to an emergency meeting of the Special Committee for National Security in Samcheong-dong.[21] The meeting was held in a cafeteria that would later become the offices of the Special

Committee. At that meeting a sub-committee was formed on which Dr. Oh Myung was asked to serve. His first responsibility as a member of this sub-committee was to provide a plan and suggest directions to help energize the nation's electronics industry.

About two months later Dr. Oh received an invitation to have dinner with Dr. Kim Jae Ik, the Chief Economic Secretary to the President. Coincidentally, both men had attended Kyunggi High School, with Kim Jae Ik preceding Dr. Oh by one year. Consequently, their relationship began on a friendly basis and with a sense of closeness. Although anecdotal, this common high school background offers a telling example of how Korean culture, including both respect for education and social relationships, helped the nation navigate a perilous time politically, economically and socially. Among those familiar with the history of Korea's telecommunications industry, the story of this meeting is widely known.

The two men met for dinner at a Korean restaurant in Gwanghwamun, near the center of Seoul. The conversation naturally flowed around Korean science and technology, with a specific focus on computers, semiconductors and the need for Korea to introduce electronic switching systems to modernize its telecommunications system. They touched on the need to begin color television broadcasting and the difficult situation Korea faced in the electronics market because it was slow to make the transition from an agricultural to an industrial society. They agreed that with the future merging of computers and telecommunications in the information society Korea would experience yet another transformation. If it was late in making this one, it seemed to them that Korea would remain perpetually a developing country.

Near the end of this long dinner meeting, which ran until curfew time, Kim Jae Ik invited Dr. Oh to come to the Blue House and work with him as a Secretary to the President for Science and Technology. Dr. Oh accepted the invitation without hesitation, and the two men worked closely together in the Blue House for the next eight months. During that period they pushed forward with color TV broadcasting, the start of privatization with the establishment of the KTA and the establishment of DACOM. Perhaps the most important outcome of their work together was the *Long-Term Plan to Foster the Electronics Sector*.

A network-centric approach: the Long-Term Plan to Foster the Electronics Sector

With the benefit of hindsight, it is clear that Korea embarked on its path of network-centric digital development in the summer and fall of 1980. Although an economist, Dr. Kim had an appreciation for the central significance of digital switching, having observed it on his visit to Taiwan some years earlier and Dr. Oh had recently completed his Ph.D. in electrical engineering at Stony Brook University. Both viewed telecommunications as the core of the electronics industry. Remarkably, their collaboration began in 1980, nearly two decades before information and communications technology (ICT) became widely recognized and referred to as an economic sector. The centerpiece of their collaboration for eight months in the Blue House was supervision of a team of experts to draft the *Long-Term Plan to Foster the Electronics Sector*.[22] The tradition of five-year economic plans had been firmly entrenched in Korea under President Park Chung Hee, and in this sense, it was a "plan-rational" and "developmental" state like Japan, in contrast to the U.S. as a "market rational" and "regulatory" state.[23]

The initial drafting of the plan took about three months in the late fall of 1980 and involved more than 20 experts representing governmental, industry and academic stakeholders in the future of Korea's electronics industry. There were representatives from five ministries (The Ministry of Trade and Industry, the Economic Planning Board, Ministry of Communications, Ministry of Finance and Ministry of Science and Technology); four companies (Samsung, Goldstar, (now LG) Anam and Sanhwa Condenser); and two research institutes (KIET and KETRI, which were consolidated in 1985 to form the Electronics and Telecommunications Research Institute, ETRI).

Dr. Hong Sung Won, who had earned his Ph.D. in electrical engineering from the University of Colorado, worked in the Blue House with Dr. Kim and Dr. Oh. The other members of the expert working group included one with a Ph.D. in electrical engineering from Purdue University, one with a master's degree in economics from Stanford and another with a master's degree from the Kennedy School at Harvard. Nearly all the remaining members of the working group had degrees from Seoul National University, Yonsei University and Korea University.[24] In other words, the working group was led by young, U.S.-trained technocrats, complemented by some of Korea's best and brightest.

The composition of the working group that wrote the long-term plan illustrates two important points. First, it was an example of public-private partnership with members from government, industry and academia. Second, the leaders and several members of the team were U.S.-educated technocrats, showing that Korea, in contrast to Japan, trusted internationally trained individuals to make high-level policy. The appointment of Dr. Oh Myung, a technocrat, as Vice Minister of Communications underscored this acceptance of internationally trained leadership and signaled a departure from the past practice of appointing only government bureaucrats to lead the ministry.[25] The leadership of such individuals ensured that Korea had the benefit of current technological expertise in implementing policies to digitize its telephone network and invigorate its electronics sector. As noted, the leaders saw the telecommunications industry and the revolutionary changes occurring in it as the future for Korea. This was a network-centric view of the sector years before it became known as the ICT sector and over a decade before Castells and other scholars addressed the rise of the "network society." In Japan, by contrast, the practice of appointing domestically educated bureaucrats persisted.[26]

The *Long-Term Plan* focused on the development of three strategic industries – semiconductors, computers and electronic switching systems. From an engineering standpoint, these three industries manufacture many of the key components used to build today's digital networks including the internet. This comprehensive, network-centric approach adopted in the *Long-Term Plan* called for the electronics sector of Korea's economy to more than double in size over a five-year period.

The significance of the 1981 *Long-Term Plan* for South Korea is difficult to overestimate, based on two of its characteristics. First, it addressed the entire electronics sector, which today is referred to more commonly as the ICT sector. It acknowledged that electronics were being shaped by innovations in digital technology and that electronic switching in telecommunications had a great deal in common with the semiconductor industry. Korea's leading technocrats realized that the nation could not have success in one without the other. The introduction of microelectronics technology into telecommunications switching in the 1970s and 1980s considerably reduced the barriers to entry into the switching industry as microelectronics have fewer moving parts, many more standardized components and are software intensive.[27] Korea seized upon this opportunity and the success of the TDX switching project in 1982 made Korea only the 10th country in the world capable of manufacturing digital switching technology.[28]

Second, the *Long-Term Plan* addressed the question of how the plan would be financed. There were two important aspects of the whole question as to how the transition in Korea's electronics sector would be financed. One was that the plan for development of the semiconductor industry (1982–1986) included an important role for Korea's large industry groups. It called for public investment of US$400 million, of which 40 percent would be financed by the National Investment Fund and the remainder by the Electronics Industry Promotion Fund, which had not yet been created. It envisaged a level of promotion ten times larger than anything attempted up to then and represented a new style of government intervention. Korea's heavy-handed HCI (Heavy and Chemical Industrialization) phase was wound back, and a new phase promoting the semiconductor industry emphasized coordination among public agencies aimed at Korea's transition to a knowledge-intensive economy.[29]

A second factor that helped solve the financial challenge was the manner in which the government reorganized the public sector telecommunications system. With the start of privatization, such companies as Samsung, Goldstar and Oriental Precision Company (OPC) were allocated profitable segments of telecoms sector in which to build up specialized businesses and were introduced to such foreign companies as ITT, Ericsson and AT&T. The profits from the telecoms activities of these Korean companies provided a secure cash flow while they were making a huge investment in semiconductor fabrication.

There were many, including government officials, who opposed the *Long-Term Plan* and even scoffed at it. They argued that the electronics industry was not a good investment because its hallmark of rapid technological change would make it difficult for Korea to catch up. Many argued that Korea should concentrate instead on strengthening labor-intensive industries. Despite such criticisms, the *Long-Term Policy* was implemented, and within five years the electronics industry had become the leading industry in Korea.

Installing a technocrat at the ministry of communications

Even with the Blue House mandate provided by the 1981 long-range plan implementation of the new policies would require the involvement of the Ministry of

Communications. The problem was that, by tradition, the Minister of Communications was a political appointee, chosen without regard for technical background or expertise with the new electronics and digital communication.

Fortunately Choi Kwang Soo, then-Minister of Communications proposed to President Chun that Dr. Oh Myung move to the Ministry emphasizing that since it "deals with technical matters it would be good if the Vice-Minister is a person who knows and thoroughly understands the technology and its significance." The President accepted his recommendation. It was also supported by Kim Jae Ik, who told Dr. Oh that he should

> Go and do the work that is needed! Truly, if we want to invigorate the electronics industry you must go to the Ministry of Communications. The core of the electronics industry is telecommunications. . . . The [Blue House] Secretariat, while overseeing these matters, will not only work smoothly, but will have a competent person in charge as Vice-Minister directly under it. Maybe it [the electronics sector] can fly like a bird.

So, although Dr. Oh Myung was leaving the Blue House, he would still work very closely with it in a new relationship that would continue throughout his seven years and seven months as Vice Minister and Minister of Communications.

The appointment of a Vice Minister with technical training directly relevant to the ongoing revolution in digital communication not only broke with the longstanding practice of making political appointments to ministerial positions. It also set a precedent that would be followed for more than a quarter century. The following year Choi Soon Dal, who received his Ph.D. in engineering at Stanford University became minister. He was followed as minister by Kim Sung Jin, who earned a Ph.D. in physics at the University of Illinois. With only a few exceptions, this new pattern of appointing technocrats, mostly U.S.-trained, to ministerial positions would continue for the next two decades and into the new millennium. A review of government records shows that 10 of the 21 ministers at the MOC and MIC after 1981 held Ph.D.s from U.S. universities and two had earned master's degrees in the U.S. In another striking pattern, 16 of these ministers had done undergraduate work at Seoul National University.[30]

"A telephone in every household" – building an equitable information society

The telecommunications policy adopted by Korea in 1980 stated the goal of universal service from the very beginning. At that time the nation had an old and completely inadequate telephone system that favored the wealthy and privileged who could circumvent the scandalous telephone service backlog. The new policy called for building an equitable information society[31] in which all Koreans, rural and urban, rich and poor would receive the same level of telecommunications service.

More specifically, the government stated the policy goal of moving quickly toward a single, fixed toll for telephone service nationwide. As of 1980, there were differential tolls to call different parts of the country, based on distance. Beyond a 100-kilometer radius the toll would increase. So, for example, it was much more expensive to place a call from Seoul to the Jeju special self-governing province than to a nearby province. Korean policymakers in the early 1980s even discussed the goal of moving toward a single toll for voice telephone service worldwide, making it one of the first countries in which such a long-range prospect was considered. In retrospect, this was farsighted considering how, just over a quarter century later, free voice and video conferencing services such as Skype have become commonplace.

The policy goal of creating an equitable information society meant that, from the start, there would be no shared use rural "party lines," such as those used in the United States in the mid-20th century. The early adoption of this goal also helps to explain why the lively net neutrality debate that took place in the United States and Europe did not gain much traction in South Korea.

Work toward implementation of the new policy began with construction of the nation's first genuinely digital network, with a fiber optic backbone and digital switches. The PSTN (public switched telephone network) was completed in June of 1987, connecting urban areas and also the majority of the nation's small fishing and farming villages. Completion of this network not only eliminated the telephone backlog in Korea, but allowed callers to receive direct international call service to 100 countries worldwide, as well as immediate local and toll-call services. The MOC provided such service to a total of 25,000 villages, each with at least 10 households, in mountainous areas and in 500 island villages as well.[32]

Completion of the PSTN in 1987 was publicly acclaimed in South Korea as ushering in the era of "one telephone per household." While not widely recognized outside of Korea at the time, it also gave the nation one of the most modern telephone networks in the world, surpassing such advanced economies as Japan whose network at that time still included a large number of older crossbar switches. June 1987 was a milestone in Korea's digital development and heralded future ICT-led development. By focusing its efforts on building the PSTN, South Korea was acknowledging the importance of networks in the information age. As Noam put it, "In the information economy, information highways are fundamental and benefit everyone. The multipliers are large for the information sector directly and for the economy as a whole indirectly."[33]

The context in which Korea began building its information superhighways included political transformation. June of 1987 not only marked completion of the nationwide PSTN but was also a decisive turning point for South Korea toward openness and democratization. What makes the shift in Korean politics during the 1980s even more remarkable is that it was partially enabled by the decision of the incumbent government to build and open access to digital communications, a move that undoubtedly contributed to more open political processes and dissent.

On June 29th Roh Tae Woo, a close associate of President Chun Doo Hwan, appeared on television and addressed the nation in his now-famous "June 29th declaration." In it he accepted, one by one, virtually all of the opposition demands, including the one for free and fair presidential elections later in the year. The speech came as a political bombshell in South Korea and the impending Seoul Olympics was one major influence on this political change. In the speech Roh commented, "At a time when the Olympics are around the corner, all of us should be responsible for preventing the national disgrace of being mocked and derided by the international community because of a division in the national consensus."[34] Although Roh was elected President in 1987, he was able to win only because the two main opposition candidates, Kim Dae Jung and Kim Young Sam, were unable to compromise and work together so they split the opposition vote.

Politically, another major impact of the highly successful 1988 Seoul Olympics was the way that the Roh Tae Woo administration utilized the event to help promote its "Northern Policy." That policy successfully established diplomatic, cultural and trade relationships with the Soviet Union, the nations of Eastern Europe, Vietnam and China, all of which had been cut off during the long Cold War era.[35]

Color television broadcasting

The story of how color television broadcasting was introduced in South Korea is an unlikely one. Throughout the 1960s and 1970s it was hard for visitors to Korea not to form a predominantly black and white image of the country. At that time Korea was indeed a war-torn, developing nation, whose mountains were lacking trees and vegetation.

In those days all of the road signs and signs on buildings were in the distinctive *hangul* script, but they were all in black and white. This was the situation through 1980. At that time, if someone had conceived of a giant electronic television screen on the side or top of a high-rise office building, there would have been no color television content for it to convey. Although 80 other nations had started color television broadcasting, South Korea was not among them.[36]

Goldstar, the predecessor of today's LG Electronics, came out with Korea's first black and white television set in 1966, through an agreement with Hitachi of Japan. In 1977, the first color TV came off a Goldstar assembly line. However, Korean companies manufactured nothing but the box, while major components inside it were imported, mostly from Japan. Moreover, even if a wealthy Korean family purchased a color television, there was no color telecasting, so they were limited to watching color videotapes.[37]

Why had the color television industry not developed? First, the government under President Park Chung Hee had forbidden both the sale of color television sets and color television broadcasting. Government restraint was a major reason that the color television industry did not develop in South Korea before 1980. At the time, it seemed that the remaining electronics industries also died together with it.

A second and related factor was the strong opinion that color television would further a sense of incompatibility between Korea's cities and its rural villages. In the 1960s and 1970s only black and white television had spread to the rural regions of Korea. Under such conditions, if color television broadcasting had begun, the citizens in rural areas would inevitably have been aware of falling behind and of their poverty and deprivation relative to urban residents. For this reason, President Park Chung Hee thought the promotion of color television would be a burden, and so he staunchly opposed it.

The very first specific task assigned to Dr. Oh Myung as a member of the Special Committee for National Security was to see that the sale of color television sets was initiated. However, even after the measure permitting the sale of color television sets took effect, they did not sell well. The reason was simple. There was no color television broadcasting.

The decision to allow color television broadcasting involved a lot of friction with other members of the special committee. Under President Park Chung Hee, the issue of color television broadcasting had been a source of dissension among citizens. It is an historical fact that, as of 1980, color television was simply not a question to be brought up for discussion. Many people held fast to that view. Some members of the special committee claimed without hesitation that color TV sales were unacceptable and broadcasting in color even more unacceptable. Both, it was argued, would foster social divisions. The issue remained highly politicized and divisive, despite the fact that 80 other countries around the world were already broadcasting in color.

Technology successes

The 1980s featured two related technology breakthroughs, each of which contributed to completion of the PSTN, Korea's first nationwide digital network. The first and most important of these was the successful TDX electronic switching project. Later in the decade another government-industry partnership led to successful development of the 4MB DRAM. In retrospect, Korea's strong global position today in broadband networks and the semiconductor industry originated with these two technology successes, both of which involved Korean-style public-private partnerships.

The switch is on: Korea's successful TDX project

Digital switching can be likened to the brain or nervous system of the internet and other modern digital networks. The switches and routers that make up digital networks provide a means of building intelligence into the network. Indeed, they make possible cloud computing, the Internet of Things, the mobile apps ecosystem, big data, artificial intelligence and blockchain technology. In other words, they are at the heart of the revolution in ICT that is transforming Korea and the world. This central role was understood by Dr. Oh Myung and Dr. Kim Jae Ik when they met for that momentous dinner meeting in 1980.

TDX, beginning with a 1982 pilot project, was the largest development project undertaken in Korea to that date. It played a crucial role in allowing South Korea to build a modern telecommunications network and extend phone and other services to citizens nationwide. It fulfilled the twin objectives of coping with the dramatically increasing demand for telephone service and developing an indigenous digital exchange technology. Also, because switching technology required sophistication in communications, computers and semiconductors, the project had a profound and synergistic effect on the entire electronics industry in South Korea. The following summary of the TDX project expands on a more detailed treatment published earlier.[38]

The individual widely credited with introducing electronic switching technology as a national priority in Korea was Dr. Kim Jae Ik. The initial public announcement of Korea's need to develop electronic switching systems was made at a social gathering in February of 1976 by the Economics Minister. After some discussion a team was formed at KIST (Korea Institute of Science and Technology), and in December of that year it drew up an "Electronic Switching System Development Plan." However, the period from about 1976 through 1981 was only a stage of basic research into electronic switching, with limited investment of about US$600,000 and approximately ten full-time staff assigned to the project over a four-year period. The low level of investment indicated that Korea was not seriously pursuing the development of electronic switching systems. Most people at that time doubted the nation's capacity to develop such systems domestically. Even among specialists, the most common reaction was to say that "in actuality this is very difficult."

As noted in Chapter 1, the timing of developments in the telecommunications switching industry worked to Korea's benefit. Production of electro-mechanical switches had required the development of a large and dedicated precision engineering capability and a high level of technical skills in manufacturing the multitude of specialized components within a switch. However, in the 1980s the switching industry saw the introduction of microelectronics with fewer moving parts and many more standardized parts. Switches became more software intensive. These changes in turn raised the social benefits for countries like Korea. Software engineers who in the past had stayed abroad to work at Bell Labs or in U.S. universities might now find employment in Korea.[39]

The Korean government's announcement of its commitment to developing electronic switch manufacturing capability was initially greeted with skepticism. ETRI, the lead organization for this effort had little prior experience in digital switching technology,[40] and there were several strong arguments against the TDX project, mainly coming from telecommunications service providers and the corporations who would manufacture the switches.

First, the service providers were naturally concerned that any domestically manufactured switches might not operate reliably and up to international standards. If they did not, from their viewpoint, which emphasized high-quality customer service, they would be better off using more expensive imported switches.

Second, companies like Samsung and Goldstar had invested heavily in foreign switching systems. They thought it would be difficult to recover their investment if an indigenous digital switching system came on line too soon.[41]

A third argument against TDX basically boiled down to concerns about Korea's technological capability to succeed. Only nine other countries in the world at that time were manufacturing electronic switches, and all of Korea's efforts during the 1970s had yet to bear fruit. Belgium was the largest exporter of such switches. BTM had a cooperative arrangement with ITT of the U.S. for export, while AT&T made switches only for the American domestic market.[42] Besides these countries, there were only Canada, Germany, Sweden, France, Japan and Britain.

Eventually several arguments in favor of the TDX development project prevailed. First, the price of foreign switches was higher than the MOC expected the price of local switching technology would be. Given the sad state of the nation's telephone network as of 1980, the local market for digital switches was large enough to guarantee that Korea would more than recover its investment in the project. Although Korea could invite competitive international bids, there seemed to be no way to reduce the price of switching systems other than developing its own production capability.

A second closely related argument was the interest in building South Korea's own technology capability. From November of 1980 through July of 1981 Korea sent 40–50 government officials and representatives of manufacturing companies to the United States to receive training on the AT&T No. 1A switch. After that training, they continued making their utmost efforts to develop a domestic manufacturing capability

Some of those who opposed the TDX project started to change their opinions in October 1981 when US$24 million[43] was allocated to start full-scale development. However, doubts persisted in some quarters. In 1982 the Telecommunications Strategy Department was established within the Ministry of Communications, and that year marked the formal start of the TDX project, which would run through December of 1995. However, even then the commitment to full-scale development was still not widely and publicly known.

As Vice Minister of Communications (MOC), Dr. Oh Myung consistently emphasized to industry and government colleagues that Korea could domestically develop electronic switching systems. Despite some risk, the benefits of the project were so sufficient that it was necessary to push forward with the development. Because ETRI had made the development plan he went so far as to prepare a document containing the names of the people from that institute who would supervise the project and then stamped the document with his official seal. In those days, one's personal or official seal carried far more weight than a handwritten signature. This stamped and signed document served as official public notice that the MOC would directly decide everything needed to manufacture electronic switches, point by point, all the way through to eventual production operations.[44]

Secretary Kim Jae Ik and other internationally educated people also had to periodically defend the electronic switching initiative. There was opposition from some very influential members of the business community at the time. However, Dr. Kim was able to blunt such criticism in part because of advice from top-notch technical experts. One of these was the MIT-trained veteran of Bell Labs, Professor Kyong Sang-Hyun of the Korea Institute of Science and Technology (KIST).

Dr. Kyong would later serve as Minister in the newly enlarged Ministry of Information and Communications under President Kim Young Sam in 1995.

Skepticism and opposition to Korea's effort to develop electronic switching technology even extended to the international arena. The presidents of certain foreign electronic switch manufacturing companies came to the Minister of Communications and asked, "Do you know how difficult the development of an electronic switching system is?" Also, they occasionally gave advice in a manner that seemed insulting. Some foreign company representatives even commented that "If Korea succeeds in domestic manufacture we'll have to reduce our international volumes." These executives didn't seem to trust what the MOC leadership said and were always trying only to market their country's brand of switching system. However, there were three international manufacturers who not only avoided such comments, but also arranged independently to assist South Korea indirectly in the development of switching systems.

As Vice Minister of Communications, Dr. Oh Myung sought out the advice of Dr. Choi Soon-dal Director of ETRI, regarding the TDX project. He initially sounded him out during a personal meeting saying, "I've just got one thing to bring up with you today. We need to push for domestic manufacturing of electronic switches. With our present technology is this possible?" The research institute director reacted somewhat skeptically saying, "Although it is relatively small, this project would require an appropriation of about $10 million."

After further discussion with Dr. Choi and colleagues at ETRI, papers were formally drawn up, with the final appropriation for TDX boldly set at US$24 million. With no objections being lodged, the Vice Minister approved South Korea's very first research and development project of such magnitude. This decision worried everyone around him. "How can you so resolutely decide to approve such a dangerous project?" they asked. The doubters questioned "How can a young, 42-year old Vice Minister with no experience make such a decision?" and "How can we later supervise the expenditure of this much money?" Such were the harsh criticisms.

However, Dr. Oh was confident that this would be an extremely easy project. His prior experience with computer development in the military convinced him that a US$24 million budget would at a minimum produce some type of electronic switch. If it could be used in the domestic market, the market would be sufficient to support it.[45] Although US$24 million was the initial investment, when all was said and done, the TDX-1 R&D budget actually, totaled US$31.6 million.[46]

The structure of the TDX project, as Korea's first major technology breakthrough in the digital era, underscored the key role of public-private partnership in the new network era. At the heart of the TDX project, ETRI was responsible for the development and control of major parts of the switching system. This involved high-level design and system integration. ETRI was assisted in its basic research by a host of public universities and other government research institutes. The basic technology was then transferred to four Korean manufacturers: Samsung, Goldstar, Daewoo and OPC. The total equipment manufacturing volume was equally divided among these four companies.

The main customer, KTA, provided all of the funds required for the project and was responsible for program management. It provided user requirements and conducted required qualification tests to ultimately commercialize the technology.

The results of the project, which had a development history of 20 years, were impressive. The number of subscriber lines increased from 10,240 with the TDX 1A to 100,000 with TDX 10.[47] Over its first decade of commercial use, Korea installed 15 million TDX lines, which accounted for more than 40 percent of the nation's telecommunications network. The bottom line was a dramatic improvement in telephone services nationwide, while saving each subscriber an estimated 40 percent of the cost for this vital service, compared with the costs if the system had continued to rely on non-Korean manufacturers.[48] The economic impact of the TDX project can also be inferred from data in Figure 2.3, showing the value and type of switches purchased by Korea Telecom between 1982 and 1993.[49]

A second and arguably more important result of the TDX project stems from the sophisticated knowledge of communications, computers and semiconductors that the project required. The TDX project paid huge dividends for many related high-tech industries including the development and manufacture of computers, consumer electronic equipment and high-tech components for a wide range of goods and services that relied on semiconductors for improved performance.[50]

A third benefit of the TDX project was the creation of intellectual property in Korea. As of 1999, over 500 patents had been registered for more than 500 TDX system components and 300 TDX software programs. Switching technology, as with other key technologies in the digital age, is a continually moving target. By the early 1990s Korean corporations had begun to develop their own versions of the TDX and to invest more in telecommunications research and development. In 1994, Yang Seung Taek, then President of ETRI, estimated that 20–25 percent of the microprocessors in the TDX-10 were imported, but the ASICs, which are critical in differentiating a product, were programmed by ETRI. Specifically, ETRI imported empty ASICs chipsets from the U.S. and UK and programmed them to create their own custom-designed microprocessors. In that sense, the firms were still very dependent on ETRI for technology. They felt that, in the newer products, such as ATM switching systems, the pace of development was too slow to keep up with global competition. To accelerate the development of new products, Samsung and Goldstar began entering into strategic partnerships with a number of overseas firms.[51]

The TDX manufacturers in South Korea also achieved success in exporting the technology. As of 1999, there were TDX exports to more than 20 foreign countries, including the Russian Federation, the Philippines, Nicaragua and Iran. The total export value of these switches was estimated at US$700 million.

A fourth important benefit of the TDX project was that it introduced quality control procedures to Korea's electronics industry. Prior to TDX, electronics products manufactured in Korea would frequently break down and require after service. The measures that were undertaken to ensure quality and reliability of the TDX switches spread to other companies and products, thereby giving an overall boost to Korea's electronics sector.

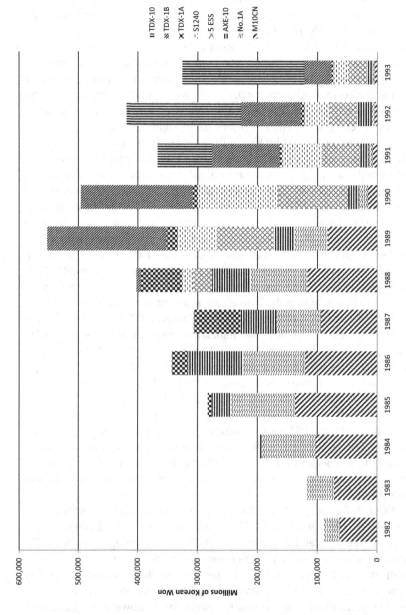

Figure 2.3 Value and type of switches purchased by Korea Telecom, 1982–1993

Source: Mytelka, L. K. (1999). The telecommunications equipment industry in Korea and Brazil. In L. K. Mytelka, *Competition, innovation and competitiveness in developing countries*. Paris: OECD Publishing, p. 139.

Finally, perhaps the most important consequence of the TDX development project was the confidence that its success instilled in everyone involved. In the end, the leadership team that developed TDX adopted the following motto: "We developed it with our brains, we made it with our hands, we teach and learn it in our language, it matches our reality and we will use it the way we wish! And one more thing, we planned the design for high product quality and safety!"

The 4MB DRAM project

The semiconductor industry is one of the world's truly global industries in that production and trade are conducted based on knowledge intensity and value-added, rather than location advantages, volume or freight charges, which dominated earlier industries. One feature of Korea's semiconductor industry is that it concentrated relentlessly, from the beginning, on memory chips. The expansion of the DRAM segment of the semiconductor market was remarkable, from introduction of the first 1K DRAM in the early 1970s to a market valued at US$40 billion by 1995. Also, the dynamics of competition in the memory chips sector are extremely demanding, with ultra-short product cycles that call for expensive investment in new process technology every two to three years. Korean companies learned to manage these product cycles.[52]

The creation of a real semiconductor industry in South Korea was a critical component of the information revolution here because of its relationship to technological progress in several other key areas. Those areas included digital switching, displays, television sets, computers, mobile phones and a host of other electronic devices.

By the late 1970s there were four private companies involved in LSI semiconductor manufacturing. However, they were still using LSI technology while the Americans and the Japanese had moved on to VLSI (Very Large Scale Integration – with between 100,000 and 1 million transistors per chip).

The promise of government support, along with direct government prodding prompted Samsung, Hyundai and Goldstar to announce in 1982 that they would make major investments in production of chips at the VLSI technology level, particularly MOS (metal oxide on silicon) memory chips such as DRAM (Dynamic Random Access Memory) chips. Although the government played an important role at this critical stage in the development of South Korea's semiconductor industry, its emergence was led by three large industry groups. Unlike the TDX project, government attempts to coordinate the efforts of major industry players were somewhat futile as each of these groups adopted a different strategy and they competed with each other to gain market share in the semiconductor industry.

In February 1982, Lee Byung Chull, the founder and chairman of Samsung famously announced that Samsung intended to become a world player in memory chip production. Moreover, he was prepared to put up 100 billion won – an astonishing US$133 million – to back his assertion. In effect, he was betting the future of the company on semiconductors.[53]

For Samsung, as for the other leading Korean companies, the road into the global semiconductor industry and mainstream chip production led through Silicon Valley. It was the location of many world-renowned semiconductor firms that employed U.S.-trained, Korean-American engineers who could be hired by the Korean *chaebol* at attractive salaries by appealing to their national pride. Silicon Valley was also home to many capital-starved start-up firms, some with excellent chip design know-how, but no manufacturing capacity. Korean companies offered these small U.S. design houses good terms for the manufacture of their chips in return for the right to license their designs. And so it was that small firms like Mosel, Vitelic and Micron and others became the source of Korea's new technology.[54]

The announcement by Samsung's Chairman Lee was followed by similar announcements from Goldstar, Daewoo and Taihan. Then Hyundai announced that it intended to enter the semiconductor field and industrial electronics generally, backed by an investment commitment of 300 billion won (US$400 million) over a five-year period, the largest commitment to that date. Hyundai's announcement led Samsung and Goldstar to announce increased commitments of their own.[55]

One estimate is that the four major players in Korea's semiconductor industry invested more than US$1.2 billion between 1983 and 1986, ten times the scale of investment in Taiwan's semiconductor industry over the same period. Furthermore, from 1983–1989 their total investment is estimated at approximately US$4 billion. The push into VLSI semiconductors not only took the industry to a new level of technological sophistication, but also to new heights of financial leverage. The investment initiative now lay with the firms themselves, more so than with the government.[56]

In December of 1983 Samsung announced that it had produced a good working version of a 64K DRAM, which was then a state-of-the art product in the semiconductor industry. However, by the time production was underway, the Americans and Japanese were already producing the next generation 256K DRAM. The *Korean firms* made Herculean efforts to establish new VLSI chip fabrication plants in 1983 and 1984, but by the time marketable products were available, the semiconductor industry was heading into a cyclical recession. Not surprisingly, early sales in the U.S. market were dismal.

These developments prompted a furious debate within government and business circles. The influential Economic Planning Board, backed by banks, argued that Korea had no future in this risky business. The Ministry of Trade and Industry, on the other hand, argued that the setbacks were cyclical, and the Ministry of Science and Technology became a major proponent of state support for Korea's long-term transition to a knowledge intensive economy.

As of 1986, the major firms in Korea's nascent semiconductor industry were still not competitive globally, and they lacked both the money and manpower to undertake the next stage of development. Also, at that time Samsung alone did not have the capacity to develop the 4MB DRAM. However, the government in a special effort to develop the new technology quickly and increase its economic impact recommended to Samsung that it work together with other companies on the 4MB DRAM.

In terms of technology development, it suggested using the method that had been successful in the TDX electronic switching project. Based on this sort of government recommendation, Samsung, Goldstar and Hyundai together achieved the development of the 4MB DRAM, and ETRI brought together teams from these three companies and coordinated the research.

Several studies of Korea's entry into the semiconductor market make it clear that the role of the government was supplementary to the already-established DRAM trajectory of the major companies involved. Nevertheless, the government played an important role. One study concluded that, although Samsung could have developed the 4MB DRAM without it, the consortium project helped Samsung to shorten its development time, an extremely important factor in the DRAM market. It narrowed its gap with the market leaders to only six months. The project was even more useful to the follower firms, Hyundai and Goldstar, who benefited from Samsung's more advanced knowledge despite a real lack of cooperation.[57]

Some individuals questioned whether Korea's efforts in the semiconductor industry could replicate the success of the TDX project.[58] Nevertheless, because the question of making good equipment for use in telecommunications involved semiconductor research, telecommunications manufacturing naturally involved managing semiconductors. Without semiconductor development, telecommunications construction would naturally run up against physical limits. Consequently, Moore's Law has had a tremendous impact not only on the semiconductor field, but on telecommunications.

Reorganization of telecommunications: privatization, policy and R&D

The first major policy change of the 1980s was a fundamental and historic restructuring of the telecommunications management system, replacing the government monopoly of the MOC with a new structure that literally created a telecommunications industry where none had existed. The new framework introduced by Dr. Oh and his colleagues consisted of a policymaking arm, a business operation arm and research and development institutes. These three areas comprised a specialized system in which each organization was assigned specific functions.

The Telecommunications Policy Office, established on January 1, 1982, assumed responsibility for formulating and promoting telecommunications policies in the public interest. It supervised and supported the public telecommunications manufacturing and construction industries. In January of 1984 the Ministry restructured the Telecommunications Policy Office into four divisions responsible for Telecommunications Planning, Telecommunications Promotion, Telecommunications Management and Information/Communications. The purpose of this reorganization was to strengthen the information and communication sectors.

The establishment of the Korea Telecommunications Authority (KTA) and the DACOM Corporation bolstered business operations in Korean telecommunications. KTA was established on January 1, 1982, to manage the telecommunications business and pursue profitability based on the public interest. The government was the sole investor, contributing 2.5 trillion won to establish the Authority. At the time, a total of 35,225 employees from 153 divisions moved from the MOC to KTA, the largest reshuffling of personnel from one organization to another in the history of the Korean government. However, there were few problems during the transition, and it set an example for the subsequent establishment of the Korea Tobacco Monopoly Corporation.

In an effort to boost the sagging data communications sector, the MOC decided that a private company needed to administer business operations in that field. Consequently, on March 29, 1982, it established the Data Communications Corporation of Korea (DACOM) under a business promotion committee on which Dr. Oh Myung served as a commissioner, as Vice Minister of Communication. In April of the same year, the first value-added telecommunications service was licensed to provide public data transmission, processing and databank services. In 1984, the Korea Mobile Telecommunications Corporation was established to oversee mobile and paging services. A year later, the Korea Port Telecommunications Corporation was formed to handle wired and wireless services relating to Korea's ports.

The formation of research and development institutes for the telecommunications sector also began during the 1980s. On March 26, 1985, the Telecommunications Research Institute of Korea was merged with the Electronics and Telecommunications Research Institute (ETRI). In February of the same year, the Institute for Communications Research (ICR) was established as a think tank for policymaking. In January of 1988 it was renamed the Korea Information Society Development Institute (KISDI). In December 1986, Dr. Kim Sung Jin was appointed as the first president of the National Computerization Agency, later renamed the National Information Society Agency (NIA) to manage and audit the government's major administrative data networks.

Deregulation of terminal suppliers (January 1981) and PSTN (1983)

Throughout the 1970s the Ministry of Communications itself was the only supplier of household telephone terminals as well as the only lender of telephone sets. This meant that subscribers did not have the option of owning a telephone or choosing which model of telephone they received. However, in January of 1981 the MOC introduced a policy that gave subscribers the opportunity to choose from a variety of options. On September 1, 1985, the MOC transferred the ownership of 2.27 million telephone sets from the KTA to telephone subscribers themselves. Consequently, manufacturers began to produce telephone sets with diverse functions and various designs, giving buyers a wider range of choices. At the same time, terminals for mobile telephones and telex machines were made available to the public, and complete liberalization of terminal distribution became a reality.

In keeping with worldwide trends in digital communications, facsimile machines, modems and personal computers began to find their way into offices and homes throughout Korea during the early 1980s. Prior to 1982 these devices could not be connected to the PSTN. Instead, they were connected to lines specifically designated for such devices. However, in March of 1983 the PSTN became available to the public and provided non-voice telecommunication services for facsimiles, modems and computers. Notably, this move preceded such deregulation of the PSTN in most European countries.

Through the 1970s private use of radio-wave frequencies had been restricted for national security reasons. In December of 1982 the MOC introduced paging service and a little later mobile communications service, initially only in metropolitan areas. The use of cordless phones was permitted beginning in September of 1983. Also, integrated management of broadcasting networks was introduced at the end of 1986, improving their operation and increasing the service coverage rate from 66 percent in 1980 to 94 percent in 1987.

Financing the transformation

The development of Korea's telecommunications sector required a tremendous investment. During the formulation of basic plans for telecommunications development in the early 1980s two measures were taken to guarantee a stable investment in telecommunications infrastructure. First, the 1981 revision of the Law on the Expansion of Information Facilities, first enacted in 1979, encouraged subscribers to purchase a fixed number of telephone bonds at the time of telephone installation. In Seoul, this cost a new subscriber 200,000 won. Second, the relatively low telephone charges at the time were adjusted to a more reasonable level. The local call rate increased from 8 won to 12 won in January 1980. It rose to 15 won in June 1981 and to 20 won in December of the same year. During the severely inflationary environment of the times, this measure was thought necessary to secure telecommunications funding.

The legacy of the 1980s telecommunications revolution

As this chapter explained, Korea's digital development originated in the 1980s. Several revolutionary breakthroughs in policy, technology and leadership left a legacy that would drive Korea's continued digital development right up to the present. That legacy includes some lessons for policymakers in developing countries and other nations around the world.

The first and perhaps most important lesson to be gleaned from Korea's experience of the 1980s is the importance of education, the basic process of the information age. It was a team led by U.S.-educated technocrats that drafted the *Long-Term Plan* and associated policies that jump-started Korea's network-centric digital development.[59] Furthermore, education in a broad sense includes citizen awareness and engagement. The "telecommunications revolution of the 1980s" led to the development of greater citizen awareness. Such awareness included convincing the nation's economic bureaucrats of the importance of telecommunications networks. The country's leading telecommunications policymakers came to believe that the free flow of information promotes the public welfare, and consequently they focused a great deal of energy on raising citizen's awareness.

A second lesson from the Korean experience is the importance of long-term planning. Digital network development requires long-range planning that takes into account new technologies and global developments in the industry. Korea took such an approach, beginning in the 1980s, with the *Long-Term Plan to Foster the Electronics Sector* and related policy initiatives.

A third lesson of the 1980s is the important role of leadership. Institutionally, the Ministry of Communications had established itself by the decade's end as one of the most powerful governmental ministries. It was affiliated with an array of new organizations, public and private, that had not existed in 1980. Both the Blue House and the country's leading economic bureaucrats now acknowledged the importance to Korea of skilled policy and technical advice in the ICT sector. Technocrats, with Dr. Oh Myung in a leading role, had triumphed over career bureaucrats. This accomplishment was only possible because of promotional leadership. [60] Starting in the 1980s, the whole government, together with the private sector and educational institutions, took on the task of promoting information culture in Korea, a pan-national effort that continues to this day. For example, although some of the key leaders were educated in electrical engineering or related technical fields, they understood the information society from political and economic perspectives as well. They believed fervently that information and communications technology would not only provide convenient services for the public, but would eventually serve as the driving force behind economic development. Dr. Oh Myung, in particular, was also confident that "genuine democratization could be achieved through the modernization of communication and the privatization of information." His dedication to spreading "information society thought" knew no bounds. Among other things, he lectured on the information society at almost every graduate school in Seoul with a specialty in this field. [61]

A fourth lesson is the value of public-private partnerships. The first such partnership was in the highly successful TDX electronic switching project. In terms of policy, over the course of the decade, the balance swung rather decisively from public to private, as the nation created new companies and began the process of privatization. The balance of policy also began to move away from monopoly toward competition. The approach to policy remained largely centralized rather than distributed. While there were many criticisms of the 5th Republic on other matters, most observers acknowledged and approved of the government's buildup of the telecommunications sector. The Blue House's Economic secretariat played a very large role in those epoch-making developments. It played the role of an orchestra conductor while each government ministry or office provided its specific services.

The evidence presented in this chapter makes it clear that Korea's remarkable digital development, in most important respects, originated in the 1980s. That decade was a crucible of political, economic and social turbulence, yet the nation made a sustained commitment to develop its electronics industry, starting with and centering on networks. Fortuitously, the 1988 Seoul Olympics gave Korea a platform to display its technological progress to the world and also to bolster the "northern policy" of opening up diplomatic, economic and cultural relationships with the former Soviet Union, China and other socialist nations including Vietnam and Eastern European countries.

The origins of Korea's digital and "ICT-driven" development are important because they illustrate the characteristic of information that economists call the "on the shoulders of giants" effect. Without the epochal developments of the 1980s it would have been nearly impossible for younger generations of engineers, scientists and government officials to achieve what they did in the 1990s and in the first decades of this new century. Not only the crucial store of information about electronic switching, semiconductors and the like was transmitted to subsequent generations, but also the confidence to carry on with a vision that inspired them! That is the story we tell in the remaining chapters of this book.

Notes

1 Cho, D. A. (2013). *Korea's stabilization policies in the 1980s*. Republic of Korea: Ministry of Strategy and Finance.
2 Korea Development Institute. (n.d.). *Social infrastructure: Establishment of the Korea institute of science and technology*. Retrieved July 10, 2016, from K-Developedia: www.kdevelopedia.org/Development-Overview/all/establishment-koreinstitute-science-technology–201412070000364.do?fldRoot=TP_ODA&subCategory=TP_ODA_SI%23.VmvD2_l96hc#.V4GmzLh952Q
3 Benedict, D. L. (1970). *Survey report on the establishment of the Korea advanced institute of science*. Daejon, Republic of Korea: U.S. Agency for International Development. Retrieved June 11, 2019 from, http://large.stanford.edu/history/kaist/docs/terman/
4 Benedict, D. L. (1970). *Survey report on the establishment of the Korea Advanced Institute of Science*. Daejon, Republic of Korea: U.S. Agency for International Development. Retrieved June 11, 2019 from, http://large.stanford.edu/history/kaist/docs/terman/
5 Hyun-kyung, K. (2015, November 16). Ex-presidential aide spearheaded shift to capital-intensive industry. *The Korea Times*.
6 Korea Joongang Daily. (2015, November 5). Remembering the good sides of Park Chung Hee. *Korea Joongang Daily*.
7 Krause, L. B. (1991). *Liberalization in the process of economic development*. Berkeley: University of California Press, p. xii.
8 Krause, L. B. (1991). *Liberalization in the process of economic development*. Berkeley: University of California Press, p. xv.
9 Krause, L. B. (1991). *Liberalization in the process of economic development*. Berkeley: University of California Press, p. xxiii.
10 Kim, K. (1991). Kim Jae Ik: His life and contributions. In L. B. Krause, *Liberalization in the process of economic development*. Berkeley: University of California Press, p. xvii.
11 The World Bank. GDP growth (annual %). Retrieved from https://data.worldbank.org/indicator/NY.GDP.MKTP.KD.ZG?locations=KR
12 Kelly, T., V. Gray and M. Minges. (2003). *Broadband Korea: Internet case study*. Geneva: ITU, p. 6.
13 Mytelka, L. K. (1999). The telecommunications equipment industry in Korea and Brazil. In L. K. Mytelka, *Competition, innovation and competitiveness in developing countries*. Paris: OECD Publishing, p. 137.

14 Sung, K. J. (1990). *Korean telecommunications policies into the 1990s*. Cambridge, MA: Harvard University Program on Information Resources Policy, p. 4.
15 Sung, K. J. (1990). *Korean telecommunications policies into the 1990s*. Cambridge, MA: Harvard University Program on Information Resources Policy, pp. 4–5.
16 Current dollars: exchange rate was approximately 600 won (Korean currency) per dollar.
17 Clifford, M. L. (1998). *Troubled tiger: Businessmen, bureaucrats, and generals in South Korea*. New York: M.E. Sharpe, p. 179.
18 Clifford, M. L. (1998). *Troubled tiger: Businessmen, bureaucrats, and generals in South Korea*. New York: M.E. Sharpe, p. 177.
19 Cho, D. A. (2013). *Korea's stabilization policies in the 1980s*. Republic of Korea: Ministry of Strategy and Finance, p. 48.
20 Cho, D. A. (2013). *Korea's stabilization policies in the 1980s*. Republic of Korea: Ministry of Strategy and Finance, p. 48.
21 Oh, M. (n.d.). History of the 1980s telecommunications revolution. Unpublished manuscript, p. 73.
22 Oh, M. and J. E. Larson. (2011). *Digital development in Korea: Building an information society*. London: Routledge, pp. 25–27.
23 Johnson, C. (1982). *MITI and the Japanese miracle: The growth of industrial policy, 1925–1975*. Stanford: Stanford University Press, pp. 17–18.
24 Jeong, H. (2006). *20 years of Korea IT policy: From the age of $1,000 to the age of $10,000*. Seoul: Korea IT News Press, p. 60.
25 Oh, M. and J. F. Larson. (2011). *Digital development in Korea: Building an information society*. London: Routledge, p. 27.
26 Johnson, C. (September 1986). *MITI, MPT and the telecom wars: How Japan makes policy for high technology*. Berkeley: Berkeley Roundtable on the International Economy.
27 Mytelka, L. K. (1999). The telecommunications equipment industry in Korea and Brazil. In L. K. Mytelka, *Competition, innovation and competitiveness in developing countries*. Paris: OECD Publishing, p. 117.
28 Oh, M. and J. F. Larson. (2011). *Digital development in Korea: Building an information society*. London: Routledge, pp. 28–35.
29 Mathews, J. A. S. (2000). *Tiger technology: The creation of a semiconductor industry in East Asia*. Cambridge: Cambridge University Press, pp. 119–120.
30 Larson, J. F. (2014). From developmental to network state: Government restructuring and ICT-led innovation in Korea. *Telecommunications Policy*, 38(4), p. 351.
31 In Korean, *chongbo bokji sahhoe*.
32 Kim, J. S. (1994). *Myung Oh: Leader of the telecommunications revolution*. Seoul: Nanam, p. 19.
33 Noam, E. (2008, November 12). A grand communication bargain. *Financial Times*. Retrieved from https://www.ft.com/content/62dbac56-b0d9-11dd-8915-0000779fd18c
34 Larson, J. F. and H. S. Park. (1993). *Global television and the politics of the Seoul Olympics*. Boulder, CO: Westview Press, p. 161.
35 Larson, J. F. and H. S. Park. (1993). *Global television and the politics of the Seoul Olympics*. Boulder, CO: Westview Press, pp. 171–187.
36 National IT Industry Promotion Agency (NIPA) History of Information and Communications, 1961–1980. Retrieved January 17, 2019, from www.nipa.kr/cyber/historySub.it?value=history_1961_9
37 LG Electronics. (2008). *LG electronics – A 50 year history* (Vol. 4). LG Electronics.
38 Oh, M. and J. F. Larson. (2011). *Digital development in Korea: Building an information society*. London: Routledge, pp. 29–38.
39 Mytelka, L. K. (1999). The telecommunications equipment industry in Korea and Brazil. In L. K. Mytelka, *Competition, innovation and competitiveness in developing countries*. Paris: OECD Publishing, p. 117.

40 *The switch is on: Korea.* (n.d.). Retrieved July 10, 2016, from http://snk5ever.free.fr/ TC-KAIST/hw1/Switch_is_on.pdf, p. 26.
41 Mytelka, L. K. (1999). The telecommunications equipment industry in Korea and Brazil. In L. K. Mytelka, *Competition, innovation and competitiveness in developing countries.* Paris: OECD Publishing, p. 140.
42 When the company was originally founded in 1920 it stood for International Telephone and Telegraph, but the company later started using just ITT. BTM was a Belgian subsidiary of ITT.
43 Over US$68 million in current dollars.
44 Oh, M. (n.d.). History of the 1980s telecommunications revolution. Unpublished manuscript, p. 181.
45 Oh, M. (n.d.). History of the 1980s telecommunications revolution. Unpublished manuscript, p. 192.
46 *The switch is on: Korea.* (n.d.). Retrieved July 10, 2016, from http://snk5ever.free.fr/ TC-KAIST/hw1/Switch_is_on.pdf, p. 24.
47 Mahlich, J. A. (2007). *Innovation and technology in Korea: Challenges of a newly advanced economy.* Heidelberg: Physica-Verlag, p. 273.
48 *The switch is on: Korea.* (n.d.). Retrieved July 10, 2016, from http://snk5ever.free.fr/ TC-KAIST/hw1/Switch_is_on.pdf, p. 25.
49 Mytelka, L. K. (1999). The telecommunications equipment industry in Korea and Brazil. In L. K. Mytelka, *Competition, innovation and competitiveness in developing countries.* Paris: OECD Publishing, p. 139.
50 *The switch is on: Korea.* (n.d.). Retrieved July 10, 2016, from http://snk5ever.free.fr/ TC-KAIST/hw1/Switch_is_on.pdf, p. 24.
51 Mytelka, L. K. (1999). The telecommunications equipment industry in Korea and Brazil. In L. K. Mytelka, *Competition, innovation and competitiveness in developing countries.* Paris: OECD Publishing, p. 144.
52 Mathews, J. A. S. (2000). *Tiger technology: The creation of a semiconductor industry in East Asia.* Cambridge: Cambridge University Press, p. 40.
53 Mathews, J. A. S. (2000). *Tiger technology: The creation of a semiconductor industry in East Asia.* Cambridge: Cambridge University Press, p. 105.
54 Mathews, J. A. S. (2000). *Tiger technology: The creation of a semiconductor industry in East Asia.* Cambridge: Cambridge University Press, pp. 121–122.
55 Mathews, J. A. S. (2000). *Tiger technology: The creation of a semiconductor industry in East Asia.* Cambridge: Cambridge University Press, p. 121.
56 Mathews, J. A. S. (2000). *Tiger technology: The creation of a semiconductor industry in East Asia.* Cambridge: Cambridge University Press, p. 126.
57 Kim, S. R. (1996). *The Korean system of innovation and the semiconductor industry: A governance perspective.* Brighton: Science Policy Research Unit/Sussex European Institute joint project, pp. 30–31.
58 Oh, M. (n.d.). History of the 1980s telecommunications revolution. Unpublished manuscript, p. 202.
59 Larson, James F. (2017). Network-centric digital development in Korea: Origins, growth and prospects. *Telecommunications Policy,* 41, p. 928.
60 Wilson, Ernest J. (2004). *III the information revolution and developing countries.* Cambridge, MA: The MIT Press, p. 94.
61 Jung-Soo, Kim. (1994). *Leader of the telecommunications revolution of the 1980s: A study on Oh, Myung.* Seoul: Nanam, p. 37.

3 Building the "information superhighways"

Government leadership in the broadband network era

This chapter brings together two topics that are central to digital development in the early years of the 21st century. One is the role of government leadership and its relationship to the private sector. The second is broadband infrastructure, one of the most important goals of government leadership and public-private partnership. Taken together, these two topics provide evidence, from Korea's experience, of what we term "network-centric digital development." They also support the argument that the decline of Korea's developmental state occurred because of its transformation into a network state.[1] The previous chapter and the first edition of this book treated developments from 1980 through 2000 in some detail, so the emphasis here is mostly on Korea's experience after the turn of the millennium.

Efforts by researchers to explain the successful use of information and communication technologies in the service of national development are full of references to the concept of leadership. As Wilson wrote,

> Without local political leadership the information revolution cannot move forward. Leaders must be willing to press changes in the face of institutional rigidity, technological backwardness and political resistance. . . . Without politics and political leadership, the information revolution simply does not occur.[2]

Leadership also requires the ability to forge an effective national public-private partnership (PPP) to achieve the expensive, difficult, long-term goals required to build an information society.

The main theme of this chapter is the role of the state, along with the private sector, in creating Korea's ICT sector as the major driver of national development and building the nation's broadband infrastructure. The main elements of developmental state theory all apply to Korea, but with several significant differences from the Japanese case. These differences all relate to Korea's transition from a developmental state to what we term a "network state."

Patterns of governance and leadership in Korea's ICT sector

Korea has a long history of centralized political power. Today that power is centered in the Blue House, the nation's presidential office and residence. Under

Korea's presidential system an election is held every five years in December, and the new president takes office in February of the following year. An exception to this timetable was the snap presidential election on May 9, 2017, following the impeachment of President Park Geun Hye. Korea's newly elected presidents routinely reshuffle the cabinet and frequently eliminate, merge or rename government ministries. On occasion there will be a sweeping government reorganization, such as the one carried out in 2008 by President Lee Myung Bak and described later in this chapter.

These regular cabinet reshuffles and changes in government structure are perplexing for international observers of ICT sector policy in Korea. As noted in a recent OECD review of industry and technology policies,

> Korea is characterized by considerable policy activism. The number of policies and programs is extremely large and their redesign and sometimes dissolution are frequent. Policies and programmes implemented abroad and considered benchmarks are regularly adopted in Korea. Indeed, few other countries create so many programmes and policies at such a pace. Furthermore, successive recent governments have ushered in significant changes in the institutions responsible for designing and implementing policy. These changes have inevitably brought a degree of disruption.[3]

The evolution of Korea's ICT sector governance

As outlined in Chapter 2, the 1980s telecommunications revolution in Korea established the basic institutional, legal and policy foundations for future growth of the nation's ICT sector. This chapter focuses on subsequent patterns in government planning and restructuring that directly affect that vital sector of the economy.

Despite the changes in government that take place with presidential elections twice every decade, viewed retrospectively over a longer period of time, there is remarkable consistency to government policies and planning for the ICT sector. The major plans approved by markedly different government administrations showed the underlying presence of a single, long-term national consensus when it came to digital technology and building the information society.

Table 3.1 shows how the structure of ICT sector governance in Korea went through several distinct historical stages corresponding to different presidential administrations. In South Korea's government structure, there are two types of institutions for ICT policy planning, coordination and implementation. The first are supra-ministerial committees that operate at the top level of government, working directly with the Blue House or the Prime Minister's office. The second type is the lead ministry, or pilot agency for coordination of industry policy, more often referred to in Korea as a "control tower."

As shown in the table, even though the MOC was increasing its influence in the ICT sector during the 1980s, the Economic Planning Board (EPB) and the Ministry of Trade, Industry and Energy (MOTIE) continued in the role of control

Table 3.1 Historical phases of ICT governance in South Korea, 1980–2017

Role of government	Developmental state		Network state		
President	Chun Doo Hwan (1980–1988)	Roh Tae Woo (1988–1993)	Kim Young Sam (1993–1998) Kim Dae Jung (1998–2003) Roh Moo Myun (2003–2008)	Lee Myung Bak (2008–2012)	Park Geun Hye (2013–2017)
Supra-ministerial leadership (Blue House, Prime Minister level)	1980–1986 – Blue House chief economic secretary	1987–1995 Information Network Supervisory Commission	1996–2007 – Informatization Promotion Committee – Prime Minister – Deputy Prime Minister (2004–2007) – Presidential e-government special committee	2008–2012 President's council on informatization strategies	2013–2017 – Blue House secretariat – Prime Minister
ICT industry pilot agency (control tower)	Economic Planning Board (EPB)	EPB, Ministry of Trade, Industry and Energy (MOTIE), MIC (from 1995)	Ministry of Information and Communication (MIC)	Ministry of Knowledge Economy (MKE)	Ministry of Science, ICT and Future Planning (MSIP)
Lead ministry	Ministry of Communications (MOC)	– MOC – MIC from 1995		Responsibilities dispersed	
Telecommunications regulation	– MOC (telecoms policy bureau) – Korea Broadcasting Commission (KBC)	– MOC – KBC	– MIC (policy bureau) – KBC	– Korea Communications Commission (KCC)	KCC

tower until that responsibility was officially transferred to the enlarged Ministry of Information and Communication by the Kim Young Sam administration in 1995. The question of where control tower responsibilities would lie was, in fact, a source of the ongoing rivalry between the MOTIE and the MIC over the years.

Leadership as a public-private partnership

From 1980 onward the Korean government enforced ICT sector policies aimed at protecting Korean companies until they were strong enough to compete with international telecom entities. The government itself initiated privatization and the introduction of competition into the marketplace, but on a basis that would allow its own companies to thrive in a growing global marketplace, rather than remaining dependent on imported electronics components. One central concern was to develop the ability, in Korea, to manufacture and export key electronics technologies.

One of the most notable changes in Korea's ICT sector over the decades examined in this book is the rising influence of the nation's large business groups. However, as Kim notes, those who argue that the growing power of Korea's large conglomerates during the postwar era came at the expense of the state's power to guide the economy use a zero-sum logic that arises "from a misunderstanding of the cooperative, but not conflict-free, nature of government and business relations" as originally set forth in developmental state theory.[4]

Reflecting on his nearly eight years of service in the 1980s as Vice Minister and Minister of Communications, Oh Myung observed that

> By its very nature the Ministry of Communications was in the position of having an inseparable relationship with large corporations. The closer one gets to industry, the more one has to prepare for friction with and criticism by industry. However, during the time I was in the Ministry of Communications, the relationship with industry was largely that of a partnership. I never saw telecommunications development as something the Ministry alone could do. Instead, I thought it was only possible in collaboration with business. I believed that sound contributions by industry were an important part of making significant forward progress in telecommunications development.[5]

During the heyday of the developmental state, from the 1970s through the early 1980s, the role of the state was so powerful that, as one of our interviewees suggested, it was the only leading actor in the telecommunications sector. That role included not only policymaking and its implementation but actually decisions about who should manufacture what and how, what to test for the testers and what to use for the operators.[6]

However, the advent of digital networks in the 1980s had a profound effect upon government and industry alike. Change in the balance of public sector versus private sector initiative in telecommunications came from both government and industry. On the government side, Korea had already begun to build governmental administration networks in the 1980s utilizing its newly completed nationwide network. These efforts took place under the auspices of the National Computerization Agency (NCA), established in 1987. The initial major project for the NCA was to build a national basic infrastructure system, essentially a nationwide network for governmental administration. That project, even many years before the introduction of the World Wide Web dramatically expanded the scope of the state's responsibilities. It forced the agency to think about future technical developments and what they would mean for government administration and services in Korea. Developments in optical transmission and electronic switching were not gradual developments to which the state could adjust. Rather, they amounted to discontinuous and revolutionary change, and the government simply could not keep up with the impact of the digital information revolution. Industry, on the other hand, was much more agile in its response to the changes brought by new digital network technologies. Compared with the governmental bureaucracy, they were capable of adjusting to the technology changes more quickly.[7]

In this manner, the role of government changed in tandem with that of industry. Both were experiencing the impact of the new digital technologies and networks. A key question is the effect of changing government structures and policies on the companies, and vice versa, along with the impact of this public-private interaction on the entire ICT sector. The question can be approached from several perspectives, as follows.

First, one needs to consider the role of government policy in the creation of its large corporate groups, their privatization and their entry into the ICT sector. For example Korea Telecom was created by the government in 1982 with establishment of the Korea Telecommunications Authority as a government owned corporation. The establishment of DACOM Corporation in 1982 and Korea Mobile Telecom in 1983 followed a similar pattern. The SK Group acquired a majority of shares in Korea Mobile Telecom in 1994 renaming it SK Telecom, the LG group acquired a majority share of DACOM in 1999, and the last government shares of KT were sold in 2002.

A second important perspective on the changing roles of government and the large corporate groups involves research and development funding and the role of government research institutes. The building of digital networks is a long-term, large-scale construction process that requires a stable and adequate funding source and the cooperation of multiple organizations. The authors of the 1981 *Long-Term Plan to Foster the Electronics Sector* recognized this and set the precedent. In 1995 the Informatization Promotion Fund was established by the enlarged and newly empowered MIC at the start of the decade-long Korea Information Infrastructure (KII) project. Managed by the MIC and based on both government budget and private sector contributions, it created a system for letting the profits from ICT fields be reallocated into the ICT sector. The fund disbursed nearly US$8 billion between 1993 and 2002 to support promotion of e-government, broadband network rollout, R&D, standardization and public education efforts. About 40 percent of the fund came from the government, 46 percent from private companies and 14 percent from miscellaneous profits and interest receipts.[8]

In quantitative terms, the dominant role of the government in Korea's R&D funding came to an end in the early 1980s. Within the ICT sector the average annual growth rate of R&D expenditure from 1993 to 2002 was 33.3 percent, and in 2001 R&D investment in ICT accounted for more than half of total R&D spending in Korea. Of this, the Korean government contributed about 10–15 percent to the total ICT-related R&D investment each year.[9]

The role of government research institutes in relation to private companies was itself a topic of heated debate in government circles during the 1980s. When these institutes were established under the Park Chung Hee administration in the 1970s their purpose was to lead industry. However, with the arrival and growth of digital networks beginning in the 1980s, the role of the large industry groups changed. Instead of simply doing what the government dictated they found themselves more often deciding what they wanted to do and then convincing the government to support it.[10]

A third important perspective on the nature of public-private interaction in Korea's ICT-led development is found in the role of the large industry conglomerates in key policy choices that shaped the ICT sector. Chronologically, the first of these decisions was the TDX switching project, in which the major industry groups followed the government lead.

The TDX project was followed by a major effort at innovation in the semiconductor industry. In the 4MB DRAM project industry groups took more of the initiative while the government played an essential supporting role. As Hong noted, there was a close fit between the demands of the DRAM industry and the structure of the Korean conglomerates. The large industry groups had access to large pools of internal capital, technological resourcefulness and manufacturing prowess, along with a management style that allowed them to enter a risky business.[11]

The initiative shown by large conglomerates in the semiconductor industry did not carry over into their role in the government's bold 1993 decision to adopt CDMA as the standard for mobile communication. The move made Korea the first nation in the world to commercialize that technology. At that time CDMA was only a theoretical concept, for which the small American company Qualcomm had patents. Although the industry conglomerates in many respects had the most to gain, as manufacturers of handsets and network equipment and also mobile service providers, it was the MOC that pushed for approval of CDMA, against the arguments of the Ministry of Trade, Industry and Energy (MOTIE), which argued that TDMA technology had a better chance of becoming commercially successful around the world. Eventually, the industry groups, led by Samsung and LG, benefited greatly from the manufacture and export of CDMA-based feature phone handsets. Ironically, their very success in this sector would contribute over a decade later to a long delay in the introduction of smartphones to the Korean market.

Finally, the relationship of government leadership and policy to the large industry groups demands a look at globalization. On the one hand, the large industry groups themselves globally expanded their research, employment and manufacturing activities, becoming globally recognized electronics brands, led by Samsung and LG. On the other hand, the weaknesses of the large industry conglomerate model were also exposed during the Asian financial crisis of 1997–1998 when several large industry groups, most notably Daewoo, declared bankruptcy.

International pressures from trade negotiations also had a big impact on Korea's ICT sector during the 1980s and 1990s. These pressures moved into a new stage with the formation of the World Trade Organization in 1995 and the conclusion of a ministerial declaration on trade in information technology products the following year. The 1990s were also remarkable for the election, successively, of two former opposition leaders as president. The first of these, Kim Young Sam, adopted *segyehwa*, which literally means "globalization," as a public slogan and pursued policies to reform Korea's economy in response to global pressures for market liberalization.

1995 restructuring: a control tower for ICT sector industrial policy

With the election President Kim Young Sam in 1994 the government reorganized by abolishing the Economic Planning Board and merging its functions with the Ministry of Finance. At the same time the Ministry of Communications (MOC) was enlarged, strengthened and renamed as the Ministry of Information and Communication (MIC). Responsibility for some of the software and information sectors was transferred from the Ministry of Trade and Industry to the new MIC. An MIT-trained Ph.D. in nuclear engineering, who had worked at Bell Labs in the U.S. before returning to Korea years earlier to work in the ICT sector, took the helm as Minister. The changes made the MIC in effect, the pilot agency for industrial policy in Korea's domestic telecommunications sector, which had become the main engine of Korea's economy. Koreans themselves, from top leaders to the general public, understood that the MIC would henceforth serve as a "control tower" for ICT sector policies.[12]

Although 1995 structural reforms formalized the role of the MIC as a control tower, they took place against a backdrop of the growing power of private companies in relation to the government. The best example of this is probably the KII project.

The government roles in this massive infrastructure project included both implementation of the public segment of the project that networked schools and public organizations and promotion of informatization to ensure widespread demand for broadband. Despite these important roles, the main reasons for the unqualified early success of the KII project were technological advances in switching and the enthusiastic response of private companies competing with each other to build out the public portion of the information superhighways.[13]

Korea's large industry groups were not quite as enthusiastic when the Korean government in 1993 made the bold decision to adopt CDMA as its mobile communications standard. The Ministry of Communication (MOC) made the decision against the objections of the Ministry of Trade, Industry and Energy making Korea the first nation in the world to commercialize CDMA-based 3G mobile technology. The six-year period between 1995 and 2002 were the strong years of

CDMA diffusion in South Korea, with Samsung and LG dominating handset sales of 3G phones domestically and also doing quite well in export markets.

The 2004 reforms: consolidating science and technology governance under a deputy prime minister

The 2004 reorganization under President Roh Moo Hyun's "participatory government" was the first to take place in the era of mobile and social media. Roh was elected with the support of young, internet-savvy members of the "386" generation-people in their 30s when the term was coined, who had attended university in the 1980s and who were born in the 1960s.

The Roh government also took office at a time of growing international awareness of the shift toward a technology-driven economy. The industrial era, in which growth could be achieved through inputs of labor and capital was reaching its limit, and some scholars predicted that technological innovation would account for 50–60 percent of future economic growth. Under these circumstances, the incoming administration of President Roh thought that the only alternative for Korea was to shift to a technology-driven economy.

To accomplish the goal of moving toward a science- and technology-oriented society with growth driven by technology and innovation, the Korean government took a three-pronged approach. First, it promoted the Minister of Science and Technology to a new Deputy Prime Minister position and appointed Dr. Oh Myung to this position. At the same time it created a new dedicated organization, the Office of Science and Technology Innovation (OSTI) under the Deputy Prime Minister for Science and Technology to oversee all microeconomic aspects of science- and technology-related industries. Microeconomic policies included research and development, industrial policy, science and technology manpower and scientific culture. In oversight of these matters the Deputy Prime Minister coordinated with the Minister of Finance and Economy who was in charge of Korea's macroeconomic policy and also held the rank of Deputy Prime Minister.

The second part of Korea's approach to becoming a technology- and innovation-driven society was the establishment of a new system to coordinate and distribute the science and technology budget. Despite quantitative expansion of investment in R&D, the nation was still not investing sufficiently in basic research, resources were concentrated in the Seoul area, and the capabilities of small and medium-sized enterprises (SMEs) were weak. To deal with this situation, a newly established National Science and Technology Council (NSTC) coordinated and allocated R&D budgets. It focused on streamlining the allocation system and enhancing professionalism in order to strategically implement national R&D projects. The NSTC emphasized alignment with national development goals, professional analysis drawing on a pool of civilian experts and incorporation of a long-term and stable perspective in the adjustment and allocation of R&D budgets.

Third and finally, the Roh Moo Hyun administration introduced a comprehensive and coordinated administrative system for use by all government departments involved with science and technology policy. The new system focused on coordinating the work of departments in the fields of science, technology and related industries, personnel and regional innovation under the guidance of the Deputy Prime Minister for Science and Technology.

As shown in Figure 3.1, the Roh administration restructuring was the first time that the Korean government elevated leadership of its science, technology and innovation policies to the level of Deputy Prime Minister. It also made the Ministry of Science and Technology (MOST) the "lead agency" but with all the key ministries reporting to the new Deputy Prime Minister. The reorganization saw two technocrats, both with Ph.D.s in electrical engineering from leading U.S. universities, assume key positions. Dr. Oh Myung assumed leadership of MOST and became Deputy Prime Minister for Science and Technology, while Chin Dae Jae headed the MIC.

MOST also took over a central role within the National Science and Technology Council (NSTC). The 2004 reform also moved responsibility for supervising the work of 3 research councils and 19 research institutions from the Prime Minister's office to MOST.[14] Consequently, MOST assumed the role of a "control tower" for innovation policy in South Korea, with authority to coordinate S&T policy, direct the GRIs, evaluate the effectiveness of policies and control S&T budgets.[15]

At the same time, to more efficiently carry out microeconomic policy, the function and authority of the National Science and Technology Council (NSTC) was strengthened. The government established an S&T-related Ministers meeting as required by law and gave the Deputy Prime Minister for Science and Technology a new role.

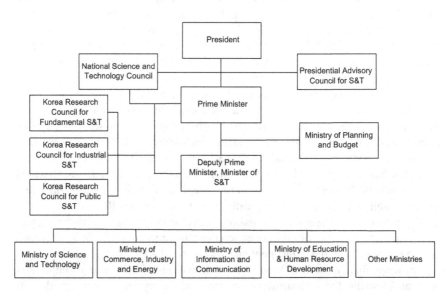

Figure 3.1 ICT sector governance in 2004

Once the new science and technology (S&T) strategy and governance structure was put in place it attracted positive interest even from overseas. James Gordon Brown, at that time the Finance Minister of Great Britain said that Korea's Deputy Prime Minister for Science and Technology would provide the most ideal test for that sort of science and technology administrative structure. The OECD, the U.S., Finland and other advanced countries also benchmarked Korea's efforts and noted its relevance for both developing and advanced economies.

The 2004 reforms represented South Korea's first major attempt to deal with innovation governance in an era of digital convergence. Convergence posed a distinct problem for countries around the world, including South Korea, which had developed their governmental regulatory schemes with the old industrial mass media in mind.[16] By placing the MIC, MOST and other key ministries under the oversight of the new Deputy Prime Minister, the 2004 restructuring explicitly acknowledged the cross-cutting nature of the policy issues posed by convergence and that science policy could not be separated from technology policy.

In 2006, Korea announced the U-Korea Master Plan, which proclaimed the ambitious goal of making the nation the world's first ubiquitous network society. Released through the Prime Minister's office and the MIC, the plan was a sure sign that the MIC had reached the zenith of its power and influence. It signaled to the world that Korea would continue its strong efforts to build next generation digital networks.

The controversial 2008 government reorganization

In January 2008 the new administration of President Lee Myung Bak announced the largest government reorganization in the nation's history, as illustrated in Figure 3.2.[17] Ostensibly aimed at increasing innovation and boosting Korea's economic growth it proved to be very controversial. Reflecting the President's own industry background as CEO of Hyundai Engineering and Construction, it suggested a much larger role for the private sector and a diminished role for government in ICT sector policy.[18]

The reorganization replaced the MIC-centered framework with a new one in which ICT would be combined with the functions of each ministry. At the ministerial level, it included the following key changes:

- The Ministry of Information and Communication (MIC) was eliminated.
- The Deputy Prime Minister for Science and Technology and the Deputy Prime Minister for Education and Human Resources Development were abolished. Also, the Ministry of Science and Technology (MOST) and the Ministry of Education and Human Resources Development (MOE) were merged to form the Ministry of Education, Science and Technology (MEST).
- The world's first Ministry of Knowledge Economy (MKE) was created, merging the former Ministry of Commerce, Industry and Energy with elements of the Ministry of Information and Communications, the Ministry of Science and Technology and the Ministry of Finance and Economy. The parts

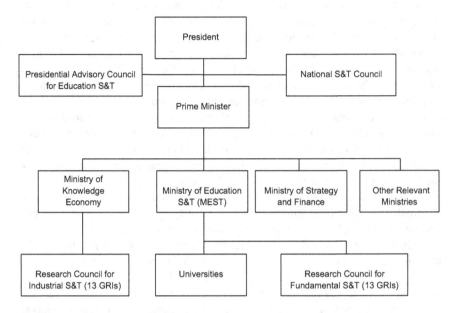

Figure 3.2 Science and technology administrative system in Korea, 2008

of MOST that were merged into the MKE included those concerned with the Daedeok Innopolis and other cluster programs.

- The Ministry of Public Administration and Security (MOPAS) assumed the MIC's former responsibilities for national informatization.
- An expanded Korea Communications Commission (KCC) was formed to deal with the convergence era and handle the core functions of the former Korean Broadcasting Commission and the telecoms policy section of the Ministry of Information and Communication.

These sweeping changes relegated science and technology (S&T) strategy to different ministries and entities throughout the whole government, much as it had been prior to the Roh Moo Hyun administration. The purpose of the Deputy Prime Minister structure established under President Roh was "to nurture creative possibilities for continued growth," and the structure generally received favorable evaluations after roughly three years of operation. However, this direction was suddenly interrupted when the Lee Myung Bak government discontinued the Deputy Prime Minister system. In addition to the abolishment of the Deputy Prime Ministers for Science and Technology and for Education and Human Resources Development, the integration of the Ministry of Science and Technology with the Ministry of Education and Human Resources Development left many people surprised and disappointed.

The 2008 reorganization ignited a public discussion and debate that involved policy circles, industry representatives, the media and, via Korea's active blogosphere, the public. It continued into the 2012 presidential election year during which it became a major campaign issue for all of the candidates. The debate on ICT sector policy included several complex issues that had, in effect, been "shelved" by the government during the fall 2007 election season. One was the question of what to do about the requirement that all mobile phones in Korea use the WIPI software protocol, a requirement that had become outmoded and was, de facto, serving as a barrier to entry of Apple's iPhone into the Korean market. Also pending were issues relating to the convergence of broadcasting and telecommunications, including IPTV, DMB and WiBRO services. The growing strength of the Chinese electronics industry also entered into the public discussion.

The following highlights from Korean press coverage illustrate the nature of the debate:

- On August 1, 2008, a leading business and economics paper carried an article that declared Korea's "once proud world's strongest IT industry has been shaken to its roots." The article claimed that only about one third of the 99 items on the KCC agenda for a meeting in late July dealt with broadcasting and communications regulatory affairs and a new IT policy agenda while all the rest were routine matters.[19]
- In October of 2008, another leading business publication carried a column by a Seoul National University public administration professor. He suggested that the reorganization had caused a "tug of war between ministries" and noted that the KCC, although charged with handling broadcasting-telecommunications convergence, was not, in fact, an IT control tower.[20]
- In December 2008 the *Financial News* published a story headlined "IT Powerhouse Korea – Developing Country's Policy." The article highlighted policy issues in three areas including (1) the WIPI requirement as a barrier to progress in the mobile sector, (2) the composition of contributions by the telecommunications industry to the broadcasting and communications development fund and (3) problems with digital contents strategy. It also noted that President-Elect Obama had recently benchmarked Korea's IT industry policy in appointing a national Chief Technology Officer in the U.S. and placing stress on high-speed internet infrastructure. It quoted a professor of business administration from Hanyang University who noted that, in the face of rapid technological change, presidential leadership was needed to "enforce policy coordination and pursue the national interest."[21]
- Also in early December of 2008, the Blue House deputy spokesperson publicly reiterated that it would not establish a separate IT control tower. She said that the "trend of the times" was to let the private sector lead the IT industry.[22]
- A May 2009 article in the *Financial News*, while acknowledging that major government ministries were all activating IT strategies, reported that this development was not welcomed by everyone. Instead, many observers voiced worries that, in the words of a well-known Korean proverb "with too many sailors the ship goes over the mountains."[23]

- In mid-2009, the CEO of KT, himself a former Minister of Information and Communications, criticized the committee-based decision-making structure of the KCC, suggesting that it was less effective as a developer of national IT strategies compared with the former MIC. He said, in part,

  The KCC was planned as a neutral, independent agency, but now the body that even has commissioners named from opposition political parties now has the power to regulate IT. Communications is part of administration and should not be a commission-based organization, and this has to be changed.[24]

- In a March 2011 interview with the *Seoul Shinmun*, Ahn Cheol Soo, then a distinguished university professor at KAIST, emphasized that the government needed an information technology control tower like the former MIC. Ahn, who would later enter the presidential race, described the weakening of Korea's IT competitiveness during the three years of the Lee Myung Bak government as "the lost three years," brought about because the government had vacated its political authority. He further suggested that Korea was unable to read such broad global trends as the Apple iPhone, social media and cloud computing mainly because the MIC, which had contributed so much, more than in other countries, to the development and commercialization of new technologies was dissolved.[25]

2013 – the creative economy initiative

During the 2012 presidential election campaign all major candidates promised stronger support for science and technology, including ICT and all acknowledged in one form or another the need for stronger government leadership and some form of control tower. Park Geun Hye, who won the presidency, campaigned on a promise to build a "creative economy."

As shown in Figure 3.3 the centerpiece of her government reorganization was the new Ministry of Science, ICT and Future Planning (MSIP).[26] The Korean name of the ministry includes the words "future," "create," and "science" and is not easily translated into English. The ministry, widely described in press and policy circles as a "superministry," had the following characteristics:

- One of the largest ministries with nearly 800 staff.
- Control of all government research and development funding (17 trillion won).
- Inclusion of all S&T-related functions that formerly had been part of MOST and were in the Ministry of Education, Science and Technology since 2008.
- Consolidation of nearly all of the functions that had been part of the former MIC.
- Telecommunications regulation was left under a downsized KCC.

The 2013 government restructuring was significant for several reasons. First, it re-introduced the control tower function that had been the aim of the 2004

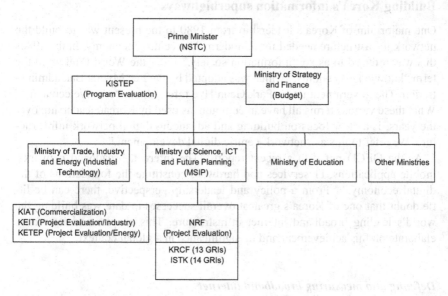

Figure 3.3 The technology and innovation system in Korea, 2013

reorganization. The new MSIP, referred to colloquially in Korea simply as the "Future Ministry," took responsibility for all R&D and industry policy aimed at building a new, creative economy.

Second, it indicated that the Park Geun Hye government took a much broader view of innovation and convergence than the prior administration of Lee Myung Bak. The Lee administration viewed digital convergence as a secondary phenomenon, with ICT supporting and subordinate to existing media and industries. After the Park Geun Hye administration publicly announced the new super ministry, some questioned the placement of ICT and S&T policy in the same Ministry, suggesting that ICT policy is short term while science and technology policy is a longer-term process. However, President-Elect Park staunchly defended her plan in the very first presidential secretariat meeting in late February, stating that "I have put convergence as the key task to revive our economy. Because the Ministry of Future Planning and Science is not approved yet, I earnestly hope the National Assembly will pass the plan as soon as possible."[27]

Third, the 2013 government reorganization underscored the continued political sensitivity of the broadcasting sector in South Korea. Approval of the new ministry was delayed for weeks, mainly because of ruling party and opposition disagreement over just where the responsibilities formerly handled by the Korea Broadcasting Commission should be placed. Some members of the opposition party claimed that the new President wanted ICT within the new Ministry in order to exert control over broadcast media.

Building Korea's information superhighways

One major aim of Korea's leadership from 1980 to the present was to build the network infrastructure needed for a modern, 21st century economy. In the 1980s this was referred to as an "information society." Later the World Bank used the term "knowledge economy," which was adopted by the Lee Myung Bak administration. The government under Park Geun Hye referred to a "creative economy." What these various terms all have in common, as used by Korean leadership over the years, is a clear focus on building and advancing digital network infrastructure. We refer to this as "network-centric digital development."

As the OECD put it in a recent report, "The Internet, broadband networks, mobile applications, IT services and hardware constitute the foundations of the digital economy."[28] From a policy and leadership perspective, there can be little doubt that one of Korea's greatest overall successes to date was building the world's leading broadband internet infrastructure. This chapter and later ones elaborate on this achievement and its significance in a global context.

Defining and measuring broadband internet

The term "broadband" is shorthand for "broad bandwidth." Bandwidth generally refers to the transmission capacity of an electronic communications device or system. In digital networks, bandwidth is usually expressed in bits per second (bps). As we use the term "broadband" in this book, it has at least three important dimensions.

First, "broadband" refers to the speed of one's connection to the internet. However, as a recent study notes, speed is only a rough measure of capacity for two reasons. One is that absolute speed, as measured in megabits per second or gigabits, does not necessarily translate into equivalent user experience or value. The second is that governments and policymakers around the world differed widely over the years in their definition of speeds required to qualify as "broadband."[29] Given continual changes in digital technologies, the term "broadband" is a moving target.

A second consideration in using the term "broadband" is availability. A broadband connection to the internet, in contrast to the old dial-up or "narrowband" connections, is always on. As distinct from speed, the always-on characteristic of broadband defines a fundamentally different user experience. Broadband, as compared to the dial-up experience that preceded it, is relatively seamlessly integrated into a user's life, at home, at the office or in the *PC Bang*. The availability of broadband networks help to explain why social networking in Korea via Cyworld preceded Facebook in the U.S. by four years. Consideration of availability also underscores why mobile communications are now the preferred mode of connecting to the internet around the world.

In addition to speed and availability *per se*, a third dimension of broadband internet is greater computing and communications power, which translates into potential value. Robert Metcalf, the inventor of Ethernet, one of today's most widely used network technologies, expressed this as a law. It states that the power of a telecommunications network increases with the number of connected users of the system. As a Berkman Center study put it, broadband internet can be defined in terms of the anticipated applications and their value rather than speed or availability alone.[30]

The diffusion of broadband globally and in Korea

Searching for information on the internet, use of e-mail and interaction through social networking sites are so commonplace today that it is easy to lose historical perspective. In fact, the World Wide Web only recently turned 20 years old, having been proposed in a 1989 research paper by Tim Berners-Lee. It is well to remember that Korea's telecommunications revolution of the 1980s, culminating in the highly successful Seoul Olympics, all took place in the pre-internet era!

Although the structure of the World Wide Web was proposed in the late 1980s, its full impact was not widely understood until the mid-1990s. The first web browser, Netscape, was only released in 1994. As of 1995 Netscape Navigator was the dominant web browser in use around the world, but by the year 2000, Microsoft's Internet Explorer had captured 80 percent of the market.

In 1994 U.S. Vice President Al Gore gave his UCLA speech calling for the construction of information superhighways. On May 26, 1995, Bill Gates sent his now-famous memorandum to Microsoft executives based on the increased speed and computing power of the internet, what we've been discussing as broadband. The subject heading of the memo was the "Internet Tidal Wave."

Even more significantly, Gates' memo was sent before the formation of Google, which only got started in January of 1996. By the mid-1990s, searching for information by looking for key words among the mounting number of pages on the internet was returning more and more irrelevant content. The founders of Google came up with a big insight to help solve this problem called the Page-Rank algorithm. Convinced that the pages with the most links to them from other highly relevant web pages must be the most relevant pages associated with the search, Larry Page and Sergei Brin tested their thesis as part of their studies and laid the foundation for their search engine. The domain name google.com was registered on September 15, 1997, and they formally incorporated their company, Google, Inc. on September 4, 1998, at a friend's garage in Menlo Park, California.

According to the founder of the World Wide Web, Google's success shows that the web needs to be understood and that it needs to be engineered.

> The web is an infrastructure of languages and protocols – a piece of engineering. The philosophy of content linking underlies the emergent properties, however. Some of these properties are desirable and therefore should be engineered in. For example, ensuring that any page can link to any other page makes the web powerful both locally and globally.[31]

On the other hand, other properties should be engineered out – such as the ability to build a site with thousands of artificial links generated by software robots for the sole intention of improving that site's search rankings – so-called link farms.

Prior to the 1990s, the internet in South Korea, as in other countries, consisted of small-scale experiments by scientists and engineers. In May 1984, DACOM began its commercial e-mail service through DACOM-net. A series of other critical events occurred in the mid-1980s that allowed Korea to meaningfully participate in the global internet. In July 1986, the first IP address (128.134.0.0) for Korea was assigned. In 1986, rules for second- and third-level domains under the. kr domain were established, and the country code top-level domain to represent Korea,. kr, was formally put in operation.[32]

South Korea adopted broadband internet on a massive scale earlier than any other nation on earth. To place Korea's broadband diffusion within the larger global context, Figure 3.4 shows Korea's performance next to two historical OECD leaders in broadband internet penetration. Although Figure 3.4 is based on penetration, a measure of subscribers per 100 population, data on usage levels

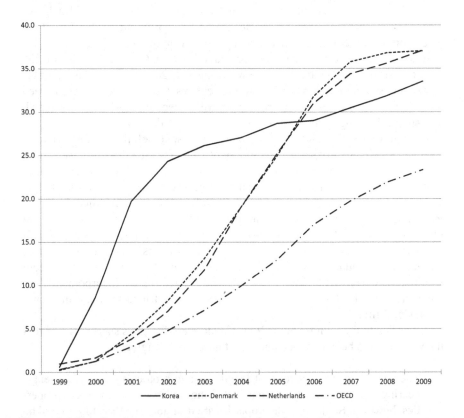

Figure 3.4 Broadband penetration in Korea and selected countries

Source: OECD.

show a similar pattern. Korea moved ahead of other nations in the world in both internet penetration and usage levels before the start of the new century. It held that lead until 2005, when several European countries, led by Denmark and the Netherlands overtook it.

Former U.S. Vice President Al Gore is well known for his advocacy of the internet, highlighted by his 1994 speech at UCLA calling attention to the need for "information superhighways." South Korea has now become well known as the first country in the world to actually build those superhighways.

Korea's answer to the ideas advocated by Gore was formally announced as the Korea Information Infrastructure (KII) plan in March of 1995. The purpose of the KII plan was to build an information superhighway that would provide advanced IT services to the public and promote informatization in every sector of society. Even more specifically and ambitiously, its purpose was to "provide various multimedia communications anywhere, anytime and to anyone, and also to turn South Korea into one of the top ten advanced countries in the IT industry by the year 2002.[33]

Several factors explain why the KII project marked a big advance over the networking projects of the 1980s. First, it involved a massive government-industry partnership. Second, it came just as the internet and the World Wide Web were becoming available globally. Third, it would provide state-of-the-art infrastructure for broadband internet approximately 4–5 years before most of the other advanced economies in the world. As one study noted, the KII project "might have been the most prominent example worldwide for governmental activities in furthering broadband deployment."[34]

The original goal of the KII project was to construct a high-speed and high-capacity "information superhighway" by the year 2015.[35] As it turned out, the project was an unqualified success and achieved all of its original goals by 2005, years ahead of schedule. There were two major reasons for early completion of the KII project. The first was continued technological improvements in switching, which will be discussed in our outline of the project. The second was the enthusiastic response of the private sector and competition in a race to build out the public portion of the information superhighway.

The mobile revolution

The mobile revolution in Korea started earlier than overall global trend and had a large impact on the nation's digital development. However, two seemingly contradictory aspects of this revolution reveal much about both the country's strengths and its relative weaknesses in building its part of the global information society. On the one hand, its hardware and networks quickly became cutting-edge technology of a sort, becoming the first nation in the world to introduce nationwide CDMA networks, mobile television and mobile Wimax (WiBRO). On the other, South Korea ironically lagged almost three years behind many other countries of the world in the actual adoption and use of mobile broadband as people in other nations were doing with the iPhone, Android phones and other smartphones. This created what has been variously referred to as the "iPhone shock," or "smartphone shock" in South Korea, beginning in December of 2009.

To place the revolution in mobile telephony in perspective, just think for a moment of the mobile handsets that were in use at the time of the 1988 Seoul Olympics. They were manufactured by Motorola and were large, heavy, brick-like hand units that in fact required two hands to operate. Today's typical mobile phones would fit within the keyboard apparatus of those bulky devices.

Since that time telephone customers around the world have "cut the cord" in a mobile communication revolution. The transformation draws much of its strength from the "law of mobility," which states that the value of a product increases with mobility. A simple measure of mobility is the percent of time that the product is available for your use.[36]

Statistically, an important milestone occurred in South Korea and worldwide, when the number of mobile phone subscriptions exceeded those for landline phones. In 1999 the number of mobile telephone subscribers in South Korea overtook the total of fixed line subscribers, as shown in Figure 3.5. This made Korea

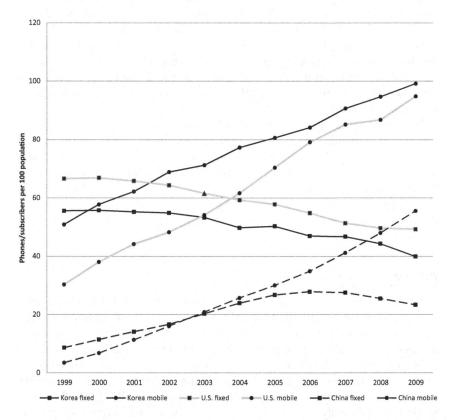

Figure 3.5 Teledensity and mobile subscribers in Korea, the U.S. and China, 1999–2009

Source: ITU.

one of the first 15 economies in the world to make this important transition.[37] Worldwide, the number of mobile subscriptions exceeded total landlines world-wide by 2001, marking a momentous year for the telecommunications industry. Underlying those numbers was the remarkable progress mobile telephony had made in reaching people who lived in the developing nations of Africa, Asia and Latin America.

As shown in Figure 3.5, Korea reached the transition point at which it had a larger number of mobile than fixed telephone service subscribers in 1999, three years before China and a full four years before the United States. However all three of these countries had rather dramatically different starting points for the ten years of data represented in Figure 3.5. The United States started with a very high teledensity of 66.6 in 1999, and a rather low mobile subscription rate of 30.24. South Korea, on the other hand, had a lower teledensity and relatively high mobile penetration in 1999, so that it was close to the transition. Finally, China started with lower numbers on both measures in 1999.

Another factor that helps to explain Korea's relatively rapid adoption of mobile communication is its decision to adopt CDMA as a national standard. The six-year period between 1995 and 2002 represented the strong years of CDMA diffusion in South Korea.[38] Although Korea was the first country in the world to commercial-ize CDMA technology and to achieve near-universal use of broadband-capable phones, the prohibitive cost of data services meant that relatively few people actu-ally used mobile broadband!

A bold decision to adopt CDMA

The mobile communications revolution in South Korea built upon the nation's experience in the 1980s in several ways. The strategic, long-term decision to develop CDMA technology took into account the future likelihood of success not only in the Korean market, but for CDMA exports. Notably, the same key people who had guided the TDX and 4MB DRAM projects in the 1980s were still in top-level decision-making positions. In short, the CDMA project was a classic example of government-led innovation.

As with the TDX project there were four main actors in the innovation system for the mobile telecommunications industry. They were the Ministry of Commu-nication, the government research institute ETRI, the local equipment manufac-turers and the mobile telecommunication service providers.

By the late 1980s, both international and corporate pressures on Korea's mobile telecommunications sector were building. One set of pressures came from continued change in digital computing and communications technologies. The other arose from bilateral trade talks with the United States and multilateral talks involving other countries urging Korea to liberalize its telecommunications market. Korean government officials had taken note that monopoly keeps prices high and encourages poor management, deterioration of service and inefficiency. Within the MOC, there was a growing sense that the ministry itself could benefit from growth in the mobile sector.[39]

The story of how the Korean government came to the decision that CDMA would be its standard for mobile telecommunications is a tour de force of the politics of telecommunications policymaking in South Korea.

The politics of liberalization in Korea's mobile telecom sector

The liberalization of Korea's cellular market came in the early 1990s and involved political struggles between industrial and bureaucratic interests and a turf war between two powerful ministries. There was also a scandal involved, but the end result of all this was a strengthened commitment to deploy CDMA. That technology choice ultimately had an enormous impact, not only on mobile communications in South Korea, but also on its exports and role in the global marketplace.

A bureaucratic turf war over control over the emerging cellular sector occurred between Korea's Ministry of Communications (MOC) and Ministry of Trade, Industry and Energy (MOTIE). The struggle revolved around three main issues: the level of industry conglomerate involvement in telecommunications services, the timing of entry for new competitors and the choice of the CDMA standard.

The introduction of cellular competitors to KMT pitted the MOC, which wanted to limit the influence of large conglomerates in the sector, against MOTIE, which was interested in boosting the manufacturing base of the conglomerates. The large industry groups had been interesting in directly operating telecommunications services for some time, but the MOC, fearing they would dominate the services market, had continually rejected their attempts to enter. The MOTIE, which oversaw high tech manufacturing and exports, disagreed, contending that the participation of industry conglomerates in telecommunications services was critical to their technological competency. The political battle between the two ministries reached a head in 1991 and reached the level of the Prime Minister's office and the powerful Economic Planning Board. The MOC tried, unsuccessfully, to engage the incumbent Democratic Liberal Party to weigh in on its side but was forced to agree to the licensing of a large industry group as the cellular competitor to KMT.

MOC and MOTIE also clashed over the timing of entry for the new mobile competitor. The MOC wanted to introduce a second carrier soon, by 1994, to encourage competition. However, MOTIE wanted to delay for one or two more years, allowing time for domestic manufacturers to develop competitiveness in infrastructure and equipment, decreasing reliance on imports. Foreign companies joined the MOC's side in this debate, and Motorola tried to alleviate MOTIE's concerns by promising it would transfer technology to Korean firms. However, the EPB weighed in on MOTIE's side. One issue that all the participants in the debate agreed upon was the strict limitation of foreign participation in the cellular market. The MOC stipulated that foreign interests could join consortia led by domestic firms, without management rights.[40]

The final issue in the debate between the MOC and the MOTIE was that of a standard for mobile telecommunications. In the early 1990s the mobile telecommunications was evolving from an analog to a digital system. Korean firms were faced with two options. They could either adapt to the new digital technology by importing the products of TDMA-based companies, or they could try to commercialize CDMA on their own.[41]

The close government-industry cooperation in the CDMA development project at ETRI was one factor that sparked a major conflict between MOC and MOTIE over the choice of CDMA as a mobile standard. By the early 1990s MOTIE's status was declining as the large industry groups gained international competitiveness and required less government support. Yet MOTIE had an institutional prerogative to retain its institutional jurisdiction over manufacturers. In 1993 it published a report critical of the MOC's plans for CDMA, arguing that TDMA, the basis behind the globally popular GSM standard, had more potential to become internationally dominant. By contrast, CDMA showed potential, but was as yet unproven. MOTIE even went so far as to launch TDMA research programs and encourage manufacturers to join.

Of course, the MOC immediately opposed MOTIE's efforts, pointing out that CDMA was technologically superior and more flexible in its future applications. The bureaucratic turf war between the MOC and MOTIE led the MOC to strengthen its plans for CDMA. It moved the date for deployment of commercial CDMA ahead by two years. More significantly, it used all of the jurisdictional authority at its disposal to promulgate CDMA as the sole domestic digital standard.

An interim outcome of the political debates was the awarding of a license in 1992 to the Daehan Telecom (Greater Korean Telecom) consortium, backed by the Sunkyong group. It was a second license for analogue mobile telephony, rather than CDMA. However, allegations of favoritism arose over President Roh Tae Woo's close relationship with the Sunkyong Group, given the government's high level of discretion in granting the license. A political firestorm ensued, forcing Daehan Telecom to return its license. The MOC was then forced to wait for a change in political leadership to conduct a second round of licensing.

In 1993, before the next round of licensing, the government announced that the standard to be adopted would be CDMA rather than the analog license that had been granted to Daehan Telecom. Thus, a long and heated debate ended with the decision to adopt the U.S.-invented CDMA as Korea's wireless standard, even though most of the world was dominated by the GSM standard. The second license was granted to Shinsegi, a consortium led by the steel company POSCO. Shinsegi had wanted to build a GSM network, since CDMA was not yet commercialized, but the MOC rejected that option.

The MOC's use of this trump card was bolstered by a 1993 decision of the U.S. Telecommunication Industry Association that it would recognize CDMA. In the end, the MOC strategy had a decisive effect on the manufacturers who found it made more sense to them than following the MOTIE. If they developed TDMA, their exports would be limited to markets in which they were newcomers, and they would be closed out of the domestic market. On the other hand, CDMA offered them access to a rapidly growing domestic market and a chance to develop high levels of competence for entering global markets.[42]

After the MOC-MOTIE debate was settled, there was a second round of research to commercialize CDMA. It was conducted largely by industry in partnership with Qualcomm, under a task force of Korea Mobile Telecom (the mobile division of KT, which had been spun off in 1988). The government subsidized approximately US$6.7 million from the Information Promotion Fund. In 1994 KMT contracted LG to provide base stations and handsets, while Shinsegi selected Samsung in 1995. These moves cemented the close carrier-manufacturer R&D relationships in the Korean mobile industry.

Development of the market in South Korea was promoted through massive entry. Two licenses for digital mobile telecommunications via CDMA technology were assigned in 1995, one to SK Telecom and the other to Shinsegi Telecom. At the same time, three PCS licenses based on CDMA technology in the 1800 MHz range were awarded to Korea Telecom Freetel (KTF), LG Telecom and Hansol. The entry of four new firms and the establishment of a nationwide standard led to very rapid expansion in the number of subscribers until 2000. Then there was a slowdown and wave of consolidation in the industry. SK Telecom merged with Shinsegi and KTF merged with Hansol.[43]

The commercial development of CDMA technology

The government decided to give the Electronics and Telecommunications Research Institute (ETRI) a lead role in developing CDMA technology. At that time, ETRI had 1,800 scientists and engineers, but most importantly, it had proven its capabilities through the successful TDX and 4MB DRAM projects in the 1980s.[44]

For ETRI, CDMA represented a major strategic direction for the lab and for its innovation capability. In fact, most of the world had already opted for GSM. However, at the time TDMA systems such as GSM and Digital AMPS were perceived to be maturing technologies that were approaching their performance limits. CDMA, on the other hand was a future technology with greater possibilities.[45] At that time, the United States telecommunications industry was wary about CDMA as an expensive, complex and unproven technology. Until the Korean government decision, CDMA existed only as a theoretical concept in which Qualcomm, a small American company, had patents.[46]

With the knowledge that they needed to move ahead into digital technology for mobile communications, Korea's Samsung and LG Electronics approached several telecom companies around the world to explore the possibility of acquiring technology. However, companies such as Motorola were only interested in exporting their products into Korea. Moreover, European firms such as Ericsson and Nokia had already developed their "global system for mobile communications" (GSM) digital technology and were not willing to share it with Korean manufacturers. This made Korea's choice to pursue CDMA easy, in a sense, because it was the only choice if the nation was to develop its own technology capacity.[47]

The MOC had outlined plans to develop a new mobile standard in 1988. The development history of CDMA in Korea was spread out over a period of nine years, beginning in 1989. The main difference between this project and the TDX project was that this was a proprietary technology originally developed and owned by the U.S. company Qualcomm. Therefore, it was conceived as a joint development project. As in the TDX project, ETRI was the main Korean institution, this time working with Qualcomm as an international partner in the technology transfer program.[48]

In 1991, four domestic manufacturers, Hyundai Electronics Industries, LG Information and Communications, Samsung Electronics and Maxon Electronics joined the project to develop a commercial CDMA system with a target date for commercial service of 1996. The key components in CDMA were three application-specific integrated circuit (ASIC) chips (MSM or Mobile Station Modem chipsets). These were initially supplied by Qualcomm, but over time ETRI and the Korean manufacturers developed their own versions. Over the course of the CDMA project, ETRI was able to keep pace with changes in technology and move to its next frontier.[49]

The total cost of the CDMA project over its entire time span has been estimated at U.S. US$65 billion. Samsung alone spent more than US$200 million on the project, which involved 1,200 researchers. Part of the financing came from service operators, who were required to donate a percentage of their revenues to research and development. Another part came directly and indirectly from consumers who had to pay a special tax on signing up, up to US$1,000 for a handset, along with deposits and activation fees.[50]

The CDMA project had a huge market creation effect for South Korea. First, its companies were able to acquire both innovation and manufacturing capability not only in CDMA but in GSM as well. Second, this new capability led to increasing exports of both handsets and base stations.

As part of the CDMA agreement, the manufacturers had to pay Qualcomm a royalty of 5.25 percent of the total handset price, excluding the cost of packing and batteries, instead of paying a royalty on only the chip and software. Over time, with the introduction of newer phone models with cameras and other features, these royalty payments became burdensome for Samsung and LG. From 1999 through 2002 they were at an estimated level of well over US$200 million annually.[51]

From 1995 to 2002 it was reported that Korean mobile handset manufacturers paid Qualcomm US$1.26 billion in royalties. Eighty percent of this amount was accounted for by royalties paid by Samsung and LG Electronics.[52] By 2007 it was estimated that Korean handset makers were paying Qualcomm more than US$500 million annually in royalties.[53] After an investigation, the Korea Fair Trade Commission (KFTC) in July of 2009 fined Qualcomm US$208 million dollars for abusing its dominant position in the market. To that date, this was the largest fine ever levied by the KFTC, Korea's anti-trust watchdog. The KFTC accused Qualcomm of collecting royalties in a discriminative way and of offering conditional rebates to Samsung and LG Electronics in return for purchasing its CDMA modem chips.[54]

In November of 2009 a development took place that appeared to signal a turning point in this dispute over royalties. Samsung Electronics and Qualcomm signed a 15-year contract for the cross-licensing of wireless telecommunications technology. Under the contract, Qualcomm gained the right to use Samsung's 57 patent licenses in mobile technology, and Samsung negotiated a reduction in the 5–5.75 percent royalty per handset it had been paying to Qualcomm. Although the details of the agreement were not made public, it was clear that Qualcomm stood ready to negotiate a similar arrangement with LG Electronics and Pantech.[55]

Despite the extremely rapid diffusion of CDMA-based digital mobile telephony in South Korea the nation was ironically one of the slower ones in the world to actually start using mobile broadband services. In fact the mobile broadband era in South Korea only took off after Apple's iPhone arrived there almost three years after its launch in the United States, and after it was already in use in more than 80 other countries around the world. In this section we look at some factors in the diffusion of mobile broadband and at why its adoption was delayed in South Korea.

The global shift from handsets to services

The introduction of Apple's iPhone in the U.S. in mid-2007 and its overwhelming success signaled the start of a revolution in mobile communications. Google's development of Android and the creation of the Open Handset Alliance were another symptom of this change. As noted earlier, telephones became more than just simply "smartphones." They were transformed into handheld computers, with internet access. That meant they could easily handle voice telephony and do so more cheaply than older technology using VOIP services like Skype.

In 2008 a review of industry trends by *The Economist* called attention to several major developments.[56] First, sales of smartphones were booming, relative to other mobile phones, and industry forecasts suggested that by 2013 they would make up 34 percent of all mobile phones, and half of the total value of the handset market worldwide. Second, as the handsets got smarter, the nature of the industry would change. It would be less about hardware and more about software, services and content, including "apps" for the iPhone, Android-based phones and their competitors. Consequently, a fierce battle had broken out among operating systems for handsets.

The existing operating systems for mobile phones, provided by Research in Motion with its Blackberry; Symbian, controlled by Nokia; and Microsoft's Windows Mobile, were all proprietary. Therefore they limited what could be done on a phone, especially as users desired more internet-related applications. The introduction of Apple's iPhone and Google's Android-based phone each in slightly different ways disrupted the status quo in the mobile telecommunications market. A general industry consensus developed that most smartphones would ultimately be powered by open source software.[57] Indeed, as of 2019 the Android mobile operating system dominates the global market.

Korea's response to the iPhone and Android

Although Korea's leading handset manufacturers, led by Samsung and LG, responded to the introduction of the iPhone by turning out a growing array of touch-screen handsets, the iPhone was nowhere to be seen in the Korean market-place. Nor for that matter was the Blackberry, a phone that was popular among business users in overseas markets. In addition to the conspicuous absence of smartphones, Korea's mobile market had the following characteristics.

First, as late as October 2009, usage of 3G mobile phones in Korea was universal, but only a little more than 10 percent of all customers purchased a data plan to use web-based services because of exorbitantly high data rates. Consumer complaints about exorbitantly high data rates were heard as late as September 2009 when SK Telecom introduced its own App Store, in a response to Apple. Users not subscribed to one of SK Telecom's fixed rate data plans would have to download apps over its 3G network at a charge of 3.5 won per kilobyte. So, downloading one of the most popular apps, the 1,349 kilobyte "2009 Pro Baseball" mobile game would cost users nearly 5,000 won for network usage, in addition to 3,000 won for the game itself.[58]

Second, the two largest mobile telecommunications service providers in South Korea, KT and SK Telecom, limited their customers to only Korean language content from the internet as part of their custom services. For SK Telecom this was NATE and for KT it was Show. This content was selected, reformatted for mobile and sold to customers. In effect it was an intranet or "walled garden." Only LG Telecom sold handsets that allowed its users to actually surf the global internet.

Third, until the spring of 2009, Korea maintained a software requirement called the wireless internet platform for interoperability (WIPI). The original idea behind WIPI was to give interoperability to mobile content providers. Prior to its introduction, SK Telecom was using its own virtual machine (VM), KTF was using Qualcomm's Brew and LG Telecom used Java. Under those circumstances content providers had to develop three separate versions of their applications. While WIPI solved that problem, it was still only a Korean standard, and it formed a barrier of sorts to entry into the Korean mobile market by Apple and Blackberry.

Korean consumers, especially the younger ones, took note of the iPhone and its cousin the iPod Touch. In the two years before the iPhone formally entered the Korean market, the iPod Touch flew off the shelves of Apple's outlets in Myong-dong, a fashionable district of central Seoul. Well over a million of these enhanced MP-3 players were sold, and many customers installed Skype on them and either used them in one of Korea's many Wi-Fi hotspots or paired them with one of Korea Telecom's eggs. An egg is a small battery-operated portable device that provides a wireless broadband (WiBRO) signal and virtually transforms the iPod Touch into a telephone, at least for those willing to use Skype.

Why was the mobile broadband era delayed?

Why did Korea, possessing some of the world's most advanced digital networks, delay its acceptance of the iPhone and therefore the introduction of the mobile broadband era? The answer to that question sheds considerable light on strategic restructuring of the nation's telecommunications sector early in the new millennium.

The first reason for this ironic situation is that Korea's major mobile service providers feared a disastrous loss of voice revenue if the market were opened up to the iPhone and other smartphones. South Korea's service providers were not alone in this concern. The soaring popularity of Skype and other voice over internet protocol (VOIP) services sent a warning signal to service providers in Europe, North America and the rest of the world. If smartphones were allowed into the marketplace, it would virtually destroy the existing business model, which relied heavily on voice revenue. As already noted, the popularity of Apple's iPod Touch among Korean youth underscored that point.[59] To protect their voice revenue, SKT reportedly insisted vigorously that Samsung Electronics and LG not include Wi-Fi capability in the mobile handsets they manufactured for SKT's services.[60] Likewise, to protect revenue from their NATE service, SKT would have little interest in opening up web surfing and web-based applications to their users.

Korea's handset manufacturers, led by Samsung and LG, were in a different position, having established themselves as major players in the international handset market. Moreover, both of these corporations were founding members of the Open Handset Alliance that backed the Android mobile OS platform. However, all of the major handset manufacturers had developed very close relationships over the years with Korea's mobile service providers. Nonetheless, Samsung and LG are both global companies with major stakes in the mobile handset business. In 2009 the business press speculated that a recent shakeup within Samsung Electronics may have been partly motivated by the impact of Apple's iPhone.[61]

Finally, the role of the government must be assessed in order to answer the question of why smartphones were late in coming to the Korean market. The answer here becomes more complex. However, it seems more than coincidental that the launch of the iPhone came during a presidential election year in South Korea and at a period when communications convergence was exerting great pressure on the policymaking process. Moreover, the newly elected government of President Lee Myung Bak eliminated the MIC, which had been the leading ministry for telecommunications policy, and the Ministry of Science and Technology. These moves were accompanied by the establishment of the Blue House-appointed KCC. Taken together, these sweeping changes gave great discretion to the private sector, both mobile service providers and handset makers, for a period of months while the new administration was being formed and preparing to pursue important policies.

Key aspects of leadership in the Korean context

This chapter examined key aspects of leadership in Korea's remarkable digital development over the past four decades, with an emphasis on broadband infrastructure. As shown in the following summary, many of them relate to major tenets of developmental state theory.

First, the Korean experience from 1980 to the present underscores the continuing utility of industrial policy as a public-private partnership, the strong theme that resonated throughout Chalmers Johnson's original formulation of developmental state theory. The size of the large industry conglomerates and their contribution to the economy changed greatly over the years, but the essence of the public-private partnership in Korea persisted. Johnson himself suggested the broader relevance of his theory to the U.S. and other developed economies.[62] More recently Noam[63] argued that, in the U.S. at least, the "government is losing its ability to do big things."

Although the role of the state has changed in the new digitally networked information environment, the Korean experience suggests that people from industry, academia, NGOs and the public still look to the government for leadership. In the era of the network state, national ICT policy leadership needs to take into account global realities and not simply local exigencies. Government restructuring, policymaking and regulation all take place in a global environment where the flow of electronic signals and services no longer respects national borders.

In the face of rapid technological change, state authorities around the world began looking at how to shift from top-down government to more decentralized governance mechanisms. These emphasized the use of partnerships and network transactions with global firms as well as the local private sector. This focus on networks and the interdependence of the state and private sector was a central and vital element of governance in Korea from 1980 onward. Dr. Oh Myung, reflecting on his career as a government minister, frequently stressed the importance of partnership with industry and compared his leadership role to that of an orchestra conductor.

Second, Korea's experience suggests a modified but still crucial role for a control tower, understood as a high-level government body responsible for industrial policy involving the ICT sector and science in the network era. The control tower metaphor suggests a high vantage point that allows a view in all directions, to communicate with and guide the planes taking off and landing. These planes, of course, come in different shapes and sizes and are independently piloted. When the government eliminated the MIC as its lead ministry in 2008, a public debate ensued in which all of the major participants – policy circles, press, private sector and public sector – agreed on the need for a return of some form of control tower. With the election of Park Geun Hye the new Ministry of Science, ICT and Future Planning (MSIP) became that control tower. Although the MSIP was a powerful ministry, its creation fell short of the Deputy Prime Minister structure for science and technology that had been introduced by the Roh Moo Hyun administration.

Third, Korea underscores the importance of a highly competent, elite bureaucracy based on a strong education system. Korea's path to ensure a competent bureaucracy had two critical features. First, it sent large numbers of its best and brightest students abroad to study, most of them at leading U.S. universities. Consequently Korea's ICT-policy leadership developed a global outlook and also became familiar with the fast-moving technology trends of the ICT industry. Second, beginning in the MOC in the early 1980s with the appointment of Dr. Oh Myung, Korea turned to technocrats with specialized training in electronics and technology fields. Some scholars refer to this as the "triumph of the technocrats." Korea also recruited many experts from the private sector who were not considered technocrats upon recruitment, but with experience came to be recognized as such.

Fourth, Korea's experience emphasizes the need for a political system that allows the bureaucracy sufficient scope to operate effectively. Otherwise put, the government bureaucracy or technocrats are insulated enough from political pressures to be effective. In Korea's case, this insulation was largely maintained over four decades and across a diverse range of presidential administrations. The Lee Myung Bak administration, with its sweeping government reorganization, represented one notable exception to this general rule.[64]

Fifth, the need for market-conforming methods of state intervention in the economy remains an imperative, but the information revolution is changing the very nature and scope of markets. Here the positive and negative aspects of Korea's large industry groups illustrate the challenge Korea faces. On the one hand these large groups, epitomized by Samsung Electronics, have become the main driver of Korea's export-based economic growth. Samsung products are now more globally recognized, purchased and used than those of many former Japanese and U.S. rivals. On the other hand, the dominance of Korean conglomerates may contribute to weakness in Korea's service sector, as vividly illustrated by the "smartphone shock" that started near the end of 2009. The arrival of the powerful handheld computers now referred to as smartphones caught Korea's government and its industry, both handset manufacturers and mobile service providers off guard. The change involved more than just a new handset, but a shift toward a smartphone-based software ecosystem in which applications, or "apps," would rapidly proliferate.

Finally, government leadership in Korea's digital development over the past four decades showed a remarkable consistency across various governmental administrations. Underlying that consistency was an approach that stressed consensus building. For example, on one occasion the MOC delayed a decision for about one year in order to assemble support for it. That was the decision to allow facsimile machines and various electronic devices to be freely connected to the newly completed PSTN. Other countries may do well to take note of the Korean experience, which suggests that government led consensus building is a vital element in policymaking and leadership for the hyperconnected era.

Notes

1 Larson, James F. and Jaemin Park. (2014). From developmental to network state: Government restructuring and ICT-led innovation in Korea. *Telecommunications Policy*, 38, pp. 344–359.
2 Wilson, E. J. (2004). *The information revolution and developing countries*. Cambridge, MA: The MIT Press, p. 13.
3 OECD. (2014, May 21). *Industry and technology policies in Korea*, p. 29. Retrieved July 31, 2016, from www.oecd.org: www.keepeek.com/Digital-Asset-Management/oecd/industry-and-services/industry-and-technology-policies-in-korea_9789264213227-en#.V51gcbh96hc
4 Kim, S. Y. (2013). The rise of East Asia's global companies. *Global Policy*, 4(2), p. 186.

5 Oh, M. (n.d.). History of the 1980s telecommunications revolution. Unpublished manuscript, p. 97.

6 Kyong, S. H. (2013, October). Former minister of information and communications. (J. F. Larson, Interviewer).

7 Kyong, S. H. (2013, October). Former minister of information and communications. (J. F. Larson, Interviewer).

8 Oh, M. and J. F. Larson. (2011). *Digital development in Korea: Building an information society.* London: Routledge, pp. 132–133.

9 Hong, D. P. (2005, May 25–26). Workshop on technology innovation and economic growth. *Workshop on technology innovation and economic growth,* Hangzhou, China, pp. 21–22.

10 Kyong, S. H. (2013, October). Former minister of information and communications. (J. F. Larson, Interviewer).

11 Hong, D. P. (2005, May 25–26). Workshop on technology innovation and economic growth. *Workshop on technology innovation and economic growth.* Hangzhou, China, p. 13.

12 Larson, James F. and Jaemin Park. (2014). From developmental to network state: Government restructuring and ICT-led innovation in Korea. *Telecommunications Policy,* 38, pp. 352–353.

13 Oh, M. and J. F. Larson. (2011). *Digital development in Korea: Building an information society.* London: Routledge, pp. 78–86.

14 Interview with Prof. Jaemin, Park. (2019, January 23). Graduate School of Technology Management, Konkuk University, Seoul.

15 Schuller, M. M. (2012). Korean innovation governance under Lee Myung-Bak: a critical analysis of governmental actors new division of labor. In J. Mahlich and W. E. Pascha, *Korean science and technology in an international perspective.* New York: Springer, p. 118.

16 Hong, S. G. (1998). The political economy of the Korean telecommunications reform. *Telecommunications Policy,* 22(8), p. 699.

17 OECD. (2009). *OECD reviews of innovation policy Korea.* Paris, France: Organization for Economic Cooperation and Development, p. 178. Retrieved June 11, 2019, from https://www.oecd.org/sti/inno/oecdreviewsofinnovationpolicykorea.htm

18 Larson, James F. and Jaemin Park. (2014). From developmental to network state: Government restructuring and ICT-led innovation in Korea. *Telecommunications Policy,* 38, pp. 354–356.

19 Anon. (2008, August 1). The Korea communications commission is an IT industry sightseer. *Maeil Kyongjae.*

20 Anon. (2008, October 22). An ill-matched IT control tower. *Korea Economic Daily (Korean language).*

21 Anon. (2008, December 10). IT powerhouse Korea – developing country's policy. *Financial News.*

22 Anon. (2008, December 7). IT control tower not needed – will not establish separately. *E Daily (Korean language).*

23 Anon. (2009, May 3). Color returning to the IT industry –but worries that "with too many sailors the ship will go over the mountains". *Financial News (Korean language).*

24 Kim, T. H. (2009, June 24). KT CEO proposes revamping telecom regulator. *The Korea Times.*

25 Anon. (2011, March 10). Distinguished KAIST professor Ahn Cheol Soo is worried about Korea's IT competitiveness. *Seoul Shinmun (Korean language).*

26 ECD. (2014, May 21). *Industry and technology policies in Korea,* p. 40. Retrieved July 31, 2016, from www.oecd.org: www.keepeek.com/Digital-Asset-Management/oecd/industry-and-services/industry-and-technology-policies-in-korea_9789264213227-en#.V51gcbh96hc

27 Ser, M. J. (2013, February 28). Park dismisses plan to raise taxes. *Korea Joongang Daily.*

28 OECD. (2015). *OECD digital economy outlook 2015*. Paris: OECD Publishing, p. 83.
29 Berkman Center for Internet and Society at Harvard University. (2010). *Next generation connectivity: A review of broadband internet transitions and policy from around the world*. Cambridge, MA: Berkman Center for Internet and Society at Harvard University, pp. 18–21.
30 Berkman Center for Internet and Society at Harvard University. (2010). *Next generation connectivity: A review of broadband internet transitions and policy from around the world*. Cambridge, MA: Berkman Center for Internet and Society at Harvard University, pp. 18–21.
31 Shadbolt, Nigel and Tim Berners-Lee. (2008, September 15). Web science: Studying the internet to protect our future. *Scientific American*.
32 Chon, Kilnam, Hyunje Park, Kyungran Kang and Youngeum Lee. (n.d.). *A brief history of the internet in Korea*.
33 Yoo, Jeong Ju, Hyeong Ho Lee and Chu Hwan Yim. (1999). *National information infrastructure in Korea*. Daejeon: Switching and Transmission Technology Laboratory, ETRI, p. 1.
34 Picot, Arnold and Christian Wernick. (2007). The role of government in broadband access. *Telecommunications Policy*, 31, p. 667.
35 *Informatization White Paper 1996*, National Computerization Agency, Republic of Korea, p. 13.
36 McGuire, Russ. (2005, December). *The law of mobility*, Sprint Nextel, p. 5. The law is named after Russ McGuire, Director of Business Strategy for Sprint/Nextel, who is credited with first stating it. As he elaborates,

> Thanks to a combination of Moore's Law, scalability resulting from Metcalfe's Law, device convergence and the increasing ubiquity of 3G wireless networks, the cost of making any product (especially one involving information) available all the time is plummeting. Therefore, just as computing power and the internet have been built into virtually every product, mobility is beginning to be built into every product.

37 *Ubiquitous Network Societies: The Case of the Republic of Korea*. ITU, April 2005, p. 18.
38 *Ubiquitous Network Societies: The Case of the Republic of Korea*. ITU, April 2005, p. 18.
39 Jho, Whasun. (2007, October). Liberalization as a development strategy: Network governance in the Korean mobile telecom market. *Governance: An International Journal of Policy, Administration and Institutions*, 20(4), pp. 638.
40 Kushida, Kenji Erik. (2008, February 1). Wireless bound and unbound: The politics shaping cellular markets in Japan and South Korea. *BRIE Working Paper 179a*, p. 26.
41 Jho, Whasun. (2007). Global political economy of technology standardization: A case of the Korean mobile telecommunications market. *Telecommunications Policy*, 31, p. 129.
42 Kushida, Kenji Erik. (2008, February 1). Wireless bound and unbound: The politics shaping cellular markets in Japan and South Korea. *BRIE Working Paper 179a*, p. 28–29.
43 Gruber, Harold. (2005). *The economics of mobile telecommunications*. Cambridge University Press, p. 141.
44 Han, In-Soo. (2007). Success of CDMA telecommunications technology in Korea: The role of the mobile triangle. In Jorg C. Malich and Werner Pascha, *Innovation and technology in Korea: Challenges of a newly advanced economy*. Physica Verlag HD, p. 290.

45 Mani, Sunil. (2005, March). Keeping pace with Globalization: Innovation capability in Korea's telecommunications equipment industry. *Working Paper 370*, India, Centre for Development Studies, p. 14.

46 Han, In-Soo. (2007). Success of CDMA telecommunications technology in Korea: The role of the mobile triangle. In Jorg C. Malich and Werner Pascha. *Innovation and Technology in Korea: Challenges of a Newly Advanced Economy*. Heidelberg, Germany: Physica Verlag HD, p. 287.

47 Dr. Hwang, Jong Sung. (2009, July 10). National information society agency NIA vice president interview.

48 Larson, James F. (1995). *The telecommunications revolution in Korea*. New York: Oxford University Press, p. 71.

49 Mani, Sunil. (2005, March). Keeping pace with globalization: Innovation capability in Korea's telecommunications equipment industry. *Working Paper 370*, India, Centre for Development Studies, p. 43.

50 Mani, Sunil. (2005, March). Keeping pace with globalization: Innovation capability in Korea's telecommunications equipment industry. *Working Paper 370*, India, Centre for Development Studies, p. 44.

51 Mani, Sunil. (2005, March). Keeping pace with globalization: Innovation capability in Korea's telecommunications equipment industry. *Working Paper 370*, India, Centre for Development Studies, pp. 45–47.

52 Handset royalties costing arm, leg. *Chosun Ilbo*. Retrieved September 18, 2003, from English.chosun.com.

53 Samsung Electronics decides to go abroad, *Chosun Ilbo*. Retrieved May 16, 2007, from English.chosun.com, http://english.chosun.com/site/data/html_dir/2007/05/16/2007051661014.html

54 Fine on Qualcomm. *The Korea Times*, July 24, 2009. Retrieved from www.koreatimes.co.kr/www/news/opinon/2009/11/202_49039.html

55 Yoo-chul, Kim. (2009, November 5). Samsung, Qualcomm renew wireless license. *The Korea Times*. Retrieved from www.koreatimes.co.kr/www/news/biz/2009/11/123_54940.html

56 The battle for the smart-phone's Seoul. *The Economist*, November 20, 2008. Retrieved from www.economist.com

57 The battle for the smart-phone's Seoul. *The Economist*, November 20, 2008. Retrieved from www.economist.com

58 Tong-hyung, Kim. (2009, September 21). SK telecom app store looks disappointing. *The Korea Times*. Retrieved from www.koreatimes.co.kr/www/news/biz/2009/09/133_52226.html

59 James F. Larson interview with Kim Shin Bae, CEO of SK C&C.

60 Tong-hyung, Kim. (2009, December 20). Wi-fi phones will slash data roaming fees. *The Korea Times*. Retrieved from http://211.234.100.245/www/news/nation/2009/12/133_57610.html

61 Moon, Ihlwan. (2009, December 15). Apple envy drives Samsung shakeup. *Business Week*. Retrieved from www.businessweek.com/globalbiz/content/dec2009/gb20091215_032027.htm

62 Johnson, C. (1982). *MITI and the Japanese miracle: The growth of industrial policy, 1925–1975*. Stanford: Stanford University Press.

63 Noam, E. (2011, March 10). The incredible shrinking U.S. broadband plan. *Financial Times*.

64 Interview with Prof. Jaemin, Park. (2019, January 23). Graduate School of Technology Management, Konkuk University, Seoul.

4   Korea's smart cities and urban information culture

Just over half a century ago Korea was still an agrarian society with the majority of its population living in rural farming and fishing villages. Today all that has changed dramatically and with it the fabric of contemporary Korean culture. However, to better understand how and why this happened and how it relates to Korea's embrace of digital technologies this chapter examines two long-term trends that occurred side by side. The first is urbanization. South Korea is one of the most highly urbanized nations in the world, having undergone so-called "compressed urbanization" in the decades following the utter devastation of the Korean War.

The second secular trend is the continued growth and evolution of digital media. Cultures, like societies, are human constructions. Although many of Korea's cultural traditions and holidays still pay homage to the nation's rural past and traditions, even a casual visitor to South Korea today cannot avoid the conclusion that Korea's culture today is a smart, technology-intensive urban one.

The McKinsey Global Institute in 2015 identified urbanization and accelerating changes in ICT in a 2015 study by the McKinsey Global Institute, urbanization as two of four long-term trends or "disruptive forces" reshaping the world. The other two were the dramatic aging of the world's population and increases in flows of trade, people, finance and data across borders. McKinsey projected that nearly half of global GDP growth between 2010 and 2025 would come from 440 cities in emerging markets, many of them as yet unknown to most Western executives.[1]

As the information revolution progresses, culture both influences the shape of a nation's media environment and is itself transformed by the new information and communication technologies. This chapter examines the two-way give and take between culture and technology in South Korea, along with the quality and texture of its information culture.

Korean urbanization in global context

Urbanization provides essential context for Korea's digital development for several reasons. First, the pace of urbanization varies greatly among nations throughout the world. Generally, developed countries experienced urbanization during or immediately following the Industrial Revolution while developing countries have encountered it more recently. Second, urbanization and economic development are interdependent processes. It was the success of macroeconomic policies that led to Korea's rapid urbanization around Seoul. However, South Korea is one of the few countries in which urbanization took place alongside sustained economic growth.[2] Third, urbanization is accompanied by important changes in the infrastructures supporting transportation and the media. In other words, urbanization affects the relative importance of what Castells calls the "space of flows" in relation to the space of places. In his words, "the new urban world arises from within the process of formation of a new society, the network society, characteristic of the information age."[3] Korea is perhaps the outstanding example in the world of the interrelationship of these processes.

Compressed urbanization

Viewed from a broad perspective, the most salient characteristic of urbanization in South Korea during the last four decades of the 20th century is its "unprecedentedly rapid" or "compressed" nature.[4] The rapidity of Korea's urbanization during the latter half of the 20th century was remarkable compared to most other developing and developed nations.

Transportation and interstate highways or expressways were a dominant metaphor of the industrial mass media era. Korea's first expressway, running from Seoul to Busan, was completed in 1970. Construction had begun in 1968 and was personally supervised by President Park Chung Hee. By comparison, construction of the interstate highway system in the U.S. was authorized over a decade earlier under President Dwight D. Eisenhower in 1956.[5] As the turn of the millennium approached, that metaphor shifted to the information superhighways of the digitally networked era, punctuated by U.S. Vice President Al Gore's speech at the UCLA Superhighway Summit about the need for America to build "information superhighways."[6]

Measuring economic development from space

More recently, researchers from Brown University discovered that satellite night lights data are a useful proxy measure for important measures of economic growth and development. They noted the conceptual problems in defining and measuring gross domestic product and the fact that GDP itself is often badly measured, especially in developing countries. Their study noted that lights data could play a key role for all countries in analyzing economic growth at sub- and supra-national levels.[7]

Figure 4.1 from the Brown University study clearly illustrates several key patterns of urbanization in Korea between the first measurement in 1992 and the second one in 2008. First, the contrast with the utter darkness of North Korea, in which both Pyeongyang and the Kaesong Industrial complex just north of the DMZ near Seoul both show up as small white specks. Second, the figure shows the massive increase of urbanization in the national capital metropolitan area

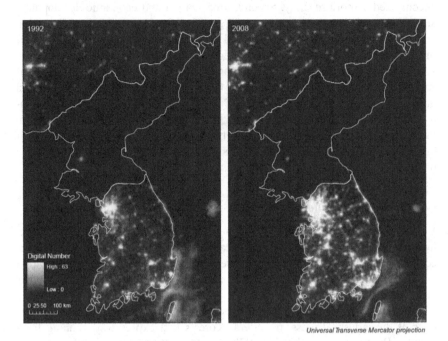

Figure 4.1 Long-term growth: Korean Peninsula

Source: Copyright American Economic Association; reproduced with permission of the American Economic Review.

surrounding Seoul and Incheon. Third, the dots running south of Seoul through Daejon to Kwangju and through Daegu to Busan in the 2008 photo are significantly enlarged. They follow the main KTX high-speed train lines from Seoul to those cities and illustrate the relationship of urban development to transportation infrastructure.

Mobile availability and speed

Today's information superhighways are increasingly mobile in nature. Measures of coverage and service quality for mobile networks in Korea and around the world are increasingly based on data gathered from mobile apps or from the telecommunications service providers. In 2016 Korea amended its Telecommunications Business Act to require service providers to release information to the public about coverage areas and levels of service.[8]

Two key measures of mobile services are download speed and availability. The latter is important because it captures one dimension of McGuire's Law, which states that the value of any product or service increases with its mobility. Furthermore, a simple measure of mobility is the percent of time a service is available.

A 2016 report by Open Signal showed that South Korea led the world in 4G availability with users able to connect to an LTE network 95.7 percent of the time. On the other key measure of 4G network speed, Korea was edged out for the top rank by Singapore, which averaged download connections of 45.9 Mbps. Korea's average download speed was 45.77 Mbps.[9]

U-Korea and smart cities

In response to rapid urbanization, many of the world's major cities, including Seoul, New York, Tokyo, Shanghai, Singapore and Amsterdam launched smart city projects and organizations such as the European Union, IBM, the ITU and many others began to explore the concept of smart cities. The assumption was that ICTs presented a viable way to update cities' traditional infrastructures to meet the needs of 21st century societies.[10] Smart cities have variously been referred to as "knowledge," "digital," "cyber" or "eco" cities. Most efforts to define and to implement the concept of smart cities include two elements. First, they focus on the use of information and communication technologies (ICTs) to address urban challenges.[11] Second, those challenges, nearly always include a reference to sustainable growth and to ensuring the wellness of citizens.[12]

In Japan, China and South Korea, interest in smart cities can be traced to the earlier broad interest in ubiquitously networked cities and societies. Accordingly the organization that as of 2016 is named the Korea Smart City Association, originated in 2005 as the U-City Forum and many of its activities continue to use the u-city rather than "smart city" terminology.[13]

The U-Korea master plan: to achieve the world's first ubiquitous society

In May of 2006 the Korean government released its U-Korea Master Plan, which made clear the ambitious character of Korea's hopes for the next stages of digital development. The Prime Minister's introduction to the plan stated that "The successful implementation of the u-Korea Master Plan, the new blueprint of Korea's informatization, will create the world's first ubiquitous society and achieve an advanced Korea."[14] The plan itself asserted that "The characteristics of ubiquitous IT – convergence, artificial intelligence and real-time – are the most effective means to upgrade the operating system of the country and to resolve the full range of social, economic and administrative issues." [15]

The plan also addressed key international issues, stating that "New social environments need to be created for a unified Korea together with the expansion of free trade based on the promotion of joint economic zones and improvement of economic cooperation between the two Koreas."[16] It noted that the technology gap with China, which had been the major export market for Korean IT products, was gradually narrowing. It also acknowledged that "The focus of the world economy is moving away from the U.S. as the Pan-Yellow Sea Rim Economic Forum has materialized and the East Asian economic zone, triggered by the rapid growth of

China, has emerged." Strategies to lead the East Asian economy, it suggested, should be formulated to prepare for the era when Asia would be the center of the world economy.

This report also explicitly acknowledged the serious threats posed by global warming and threats to the natural environment. In underscoring this point, it noted that "Intensive environmental monitoring and regulating lead to high added value for environmental industries.[17]

The report mentioned a growing gap in the amount spent on education between high and low income consumers and suggested that this was one factor contributing to conflicts between regions, social classes and generations. Finally, it noted that Korea's society was aging more rapidly than other advanced economies. As Korea becomes a "super aged society" welfare for the elderly, contraction of the economically active population and inter-generational conflicts might become social and economic problems.

The u-cities movement

In February of 2006, the Ministry of Information and Communication and the Ministry of Construction and Transportation signed an MOU on the u-city project. That project was aimed at building industry-wide partnerships between the high-tech and construction sectors to integrate advanced IT infrastructure into the construction of sustainable cities. Under the MOU, the two ministries agreed to cooperate in such areas as:

- enactment of regulations for the construction of u-cities,
- development and certification of a standardized u-city model,
- promotion of u-city pilot projects,
- R&D development of u-city-related technologies, and
- discovery and promotion of u-city-related subjects.

In addition, the two ministries agreed to exchange information and personnel and to undertake international activities for global standardization and advancement into international markets. Under the MOU, they agreed that Korea would push ahead with its nationwide u-cities plan. Several city governments, including Seoul, Busan and Incheon, expressed their intent to independently pursue u-city development, and all six regions in Korea had plans to invest in their own u-city projects.[18] In 2007 Korea passed a law on the construction of u-cities, allowing the central government to have some policy influence on the diverse local efforts.[19]

New Songdo: Korea's brand new ubiquitous city

New Songdo International City in Incheon aimed to be the world's first entirely new ubiquitous city, built from scratch. Rising up on almost 1,500 acres of reclaimed land off the coast of Incheon, it is a one of the largest planned city projects ever undertaken anywhere in the world.

New Songdo assumes special prominence among Korea's u-city projects for several reasons. First, it is an integral part of the Incheon free economic zone (IFEZ), which includes Korea's new Incheon International Airport. Created in 2003, IFEZ is the largest and most important of the nation's eight Free Economic Zones. From its inception a goal of IFEZ was to make Incheon a major regional and international hub for (1) communications (2) sea and land transportation and (3) air transportation. Measured in square kilometers, the entire zone is about 2.5 times the size of Manhattan.[20] Second, the airport and IFEZ are the closest ones to South Korea's big cooperative industrial project with North Korea at Kaesong. The longer-term vision of Korea as a hub in Northeast Asia contemplates a unified Korea. Third, as noted already, New Songdo City, unlike existing cities, presents an opportunity to incorporate ambient intelligence into the city from scratch. Finally, New Songdo City has particularly ambitious goals to become a global educational and research hub for the 21st century.

Education in New Songdo

No other area underscores the ambitious goals of the Songdo International City development better than its plans for education. The Korean government initially invited ten international universities, mostly from the United States, to set up branch campuses as part of a single global campus in Songdo. To emphasize the seriousness of its invitation, it offered approximately US$1 million to each school for planning purposes. Originally named the Songdo Global University Campus, in 2015 the name was changed to Incheon Global Campus.

Songdo received a significant boost as an education hub in 2006 when Yonsei University reached a final agreement with the Incheon Metropolitan City Government to build a new global academic complex in Songdo. The first phase of the complex included a 228-acre campus with a residence college for freshmen and a university global village, intended to improve the living environment for distinguished foreign scholars. Yonsei University was traditionally Korea's leader in international studies, being the first to develop a large international division and the first to create a graduate school of international studies. Also in 2006 Yonsei launched the Underwood International College, the most significant commitment by a Korean university to date to a four-year, English-only undergraduate program with the aim of attracting students from around the world and producing global leaders.

In December 2008 Stony Brook University (SBU), part of the State University of New York (SUNY) system, announced plans for a branch campus in Songdo. It signed a fund support agreement with the Incheon FEZ[21] that would eventually lead to the founding of SUNY Korea. Dr. Oh Myung is the founder of SUNY Korea and also its honorary president.

To surmount the many legal, administrative and other hurdles in making the vision of SUNY Korea a reality, Dr. Oh worked hand in hand with Dr. Yacov Shamash, longtime Dean of Stony Brook University's College of Engineering and Applied Sciences and, since 2000, Vice President for Economic development at Stony Brook University. Although many others were involved, these two individuals led the efforts to establish SUNY Korea.

SUNY Korea opened its doors on the Incheon Global Campus in March 2012, becoming the first American university to open a campus in Korea. James F. Larson served as provost of SUNY Korea and chaired its Department of Technology and Society. As of this writing, SUNY Korea offers undergraduate and graduate programs from five SBU departments, including Computer Science, Technology and Society, Mechanical Engineering, Applied Mathematics and Statistics and Business Management. It also offers two signature programs of the Fashion Institute of Technology (FIT), Fashion Design and Fashion Business Management.

Students in SUNY Korea's SBU departments all spend at least one full academic year on the Stony Brook University campus in New York. Applications and admissions to SUNY Korea continue to grow at a healthy rate, adding momentum to this groundbreaking effort to introduce a new paradigm for educating global leaders.

Smart Seoul

Smart city projects around the world generally divide into two types. Greenfield smart cities are those built from scratch, while brownfield cities consist of efforts to make existing cities smart. Both focus on the building or renovation of ICT infrastructure, for use not only in cyberspace, but also to communicate among elements of physical infrastructure, transmitting real-time data on a city's status by way of sensors and processors in real-world infrastructure.[22] While Songdo is Korea's leading example of a greenfield smart city, Seoul is undoubtedly its primary example of the brownfield type.

Smart Seoul is built on the following three pillars that also constitute essential traits of all smart cities:

- ICT Infrastructure: Securing next generation ICT infrastructure is critical to the success of emerging smart city services. This requires anticipating future service demands.
- Integrated City-Management Framework: The many integrated subsystems, meta-systems and individual, building-block systems of a smart city will work in harmony only through the strictest adherence to common standards.
- Smart Users: ICTs are the tools to enable a smart city, but are of no use without smart-tech users able to interact with smart services. Increasing access to smart devices and education on their use, across income levels and age groups, must remain one of a smart city's highest priorities.[23]

The character of Korea's urban information culture

In South Korea, the interaction between humans and technology since the 1980s produced an information culture that is very technology-friendly and distinctly urban in character. The following pages offer a qualitative and selective review of some salient characteristics of that urban information culture.

Seoul's screens

The Seoul metropolitan area, home to half of South Korea's population, is today a city with a profusion of bright, colorful digital screens. This was not always the case. Well into the 1970s electric billboards and lighted signs were prohibited by law, as was color television. The digital revolution of the 1980s changed all of that, spurring growth in the graphics industry and allowing Korea's venerable alphabet, *hangul*, to start appearing on signboards in a never-ending array of creative new fonts.

Today, Seoul's digital screens range in size from the façade of an entire building to the screens on mobile phones that everyone carries. There are small TV screens in subways, in elevators and on the navigation devices that virtually all taxis and private automobiles use. During the design-focused "2009 Seoul Festival of Light" the city lit the largest LED screen in the world for the first time. Located on the façade of the former Daewoo Building opposite Seoul Station, the screen was 99 meters wide and 78 meters high. It consisted of 42,000 LEDs covering the face of the 19-story building and its initial display included "Walking People," a work by the renowned British pop artist Julian Opie.[24]

Seoul is also a city of rooms (*bangs*) many of which prominently feature electronic screens. Korea's room culture includes the *tabang* (tea room) culture, which still exists but for many people has been eclipsed by the rise of Starbucks and a host of other coffee house chains. Other types of rooms include the PC Bang (PC Room), *norae*-bang (Karaoke) and the *jimjil*-bang. Culturally speaking, Koreans think of a bang as a multifunctional space, whose purpose changes according to the occupant's will. Thus, *jimjil bangs* contain sauna-like rooms, baths, sleeping rooms, snack bars and PC Rooms.

The PC Bangs, online games and the birth of e-sports

For the past two decades South Korea has been a leader in the development of a global sports culture around electronic-sports, more commonly referred to as e-sports. Their development has had a significant impact on Korean popular culture and by extension the shape of cyberspace around the world. The story of how and why this happened involves the Korea Information Infrastructure (KII) plan to build broadband infrastructure, the introduction and rapid spread of PC Rooms and the popularity of a massive multiplayer online game called StarCraft.

As shown in Figure 4.2, the number of PC Bangs in the country increased more than sevenfold from 1998 to 2001 when it peaked at 22,548. Within three years of their introduction, they had essentially reached saturation levels, which meant that one could find a PC Room in almost every urban neighborhood in the country!

There were several reasons for the explosive growth of PC Rooms. First, PC Rooms, also referred to as internet cafés, played an important role in the rapid diffusion of broadband internet in Korea. When they arrived on the urban scene few Koreans had high-speed internet access at home or at the office. At that stage of network development in Korea they represented a solution to the "last mile"

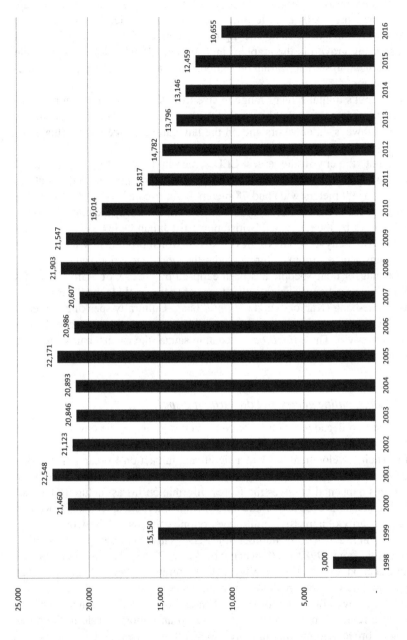

Figure 4.2 Number of PC Bangs in Korea, 1998–2016

Source: Korea Creative Content Agency white papers on Korean Games.

problem, referring to the challenge of building the final leg of network infrastructure to deliver telecommunications services to end users.[25]

When the massive Asian Financial Crisis struck near the end of 1997, the Kim Dae Jung government temporarily suspended the "community access centers" that had been established by the Kim Young Sam administration. These were public spaces throughout the country where citizens could access the internet for free. Some of the workers who had been laid off from these centers turned to more entrepreneurial efforts, including the creation of PC Bangs, which gave customers a computer terminal and online access for a small hourly fee. In this way, PC Bangs filled a void left after the closing of community access centers and became a strong force in promoting the usefulness and convenience of the internet.[26]

In July of 1998 Thrunet began deploying broadband service using cable modems for only US$25 per month. This move came as debates in government and industry about the proper level of service fees continued. Given such competition from Thrunet, Hanaro Telecom could no longer delay entering the market. In April 1999 the company commenced broadband service, offering both DSL and cable, leasing cable capacity from Powercomm, a subsidiary of KEPCO and KT. It matched Thrunet's price of US$25 per month. Furthermore, Hanaro bundled broadband with its basic telephone service and offered free installation, at only about US$40 per month. With such aggressive pricing, it acquired more than a million subscribers within 18 months. Hanaro's market entrance is often considered the start of South Korea's broadband explosion. The introduction of low, flat-rate ISP pricing for broadband internet contributed greatly to the popularity of StarCraft and other online games.

Second, StarCraft was introduced to the online game market by American game developer Blizzard Entertainment in 1998, boosting demand for high-speed broadband as opposed to slower dial-up connections to the internet.[27] Multiplayer online games require speed in order to give a realistic sensation of simultaneous interaction online with many other players. Korea accounted for one third of StarCraft's total global sales in 1998, mostly to PC Rooms.[28]

Star Craft became immensely popular among middle and high school students, who came home late at night after playing the game in PC Bangs. Hanaro Telecom picked up on this phenomenon and focused its advertising on the fact that, with its ADSL service, they could play the game at home. This appeal to parents was so successful that the waiting list for Hanaro Telecom's ADSL service reached 500,000 and stayed at that level for a long time.

Third, the PC Bangs filled a need that fit almost perfectly with Korean culture and traditions. Research by Chee found that PC Bangs functioned as "third places" or places of psychological comfort and support for young people in Korea.

> At a third place, such as a PC bang, one can choose from online games, e-mail, online chat, web surfing, visiting matchmaking sites, people watching, eating, smoking being with big groups of friends or just being with one's significant other in a friendlier setting.[29]

Some of her respondents were students who used PC Bangs as a warm and cheap place to meet during the cold Korean winters.

Finally, the steady decline in the number of PC Rooms beginning in 2010 can be attributed largely to the arrival of smartphones, both in Korea and globally. The mobile game platform emerged as one of the most important segments of Korea's game industry, and some of this growth came at the expense of the PC Bang market.

Size and growth of the game market

When StarCraft first arrived in the Korean market it was played mainly by young men in PC Bangs. In the early years, owners of the PC Rooms noticed that people were not only coming to play StarCraft, but that many were also coming to watch others play. In December 1998 the Korea Professional Gamers' League (KPGL) was formed, but it was in 1999 when cable television channels began broadcasting matches regularly that StarCraft, and by extension e-sports, became a mainstream cultural phenomenon. A liaison officer for the Korea e-sports Association, successor to the KPGL, explained the popularity of StarCraft in Korea in terms of its alignment with Korea's *gukminseong* or national character. "Koreans are very fast, very intelligent, and want to talk with others. And so StarCraft was very good for Koreans."[30]

In 2004 South Korea even established a Game Science High School. Students who were accepted into the school completed all of their national requirements for high school graduation the first year and then worked on industry-related projects for the remaining three years of high school.[31]

Korean e-sports rapidly developed a global following, as Korean StarCraft pro gamers dominated international competitions. Their success led to the first World Cyber Games (WCG) an Olympics-style e-sports competition organized by Samsung, the Ministry of Culture Sports and Tourism and the Ministry of Information and Communication, first held in Seoul in 2000. President Kim Dae Jung addressed the international audience in a pre-recorded video, saying, "I hope that the first WCG will help our nation to become recognized as one of the leaders in games, knowledge industry, and IT infrastructure, as well as help the world's game-loving young people exchange information and build friendships."[32]

As of 2017, South Korea was the world's third largest market for online video games and its fourth largest mobile game market. The three-largest markets for mobile games in 2017 were China (US$17.3 billion), Japan (US$8.56 billion) and the U.S. (US$7.28 billion).[33] Three sectors, online games, mobile games and PC Bangs make up over 95 percent of the market. As of 2016 online games accounted for 43.6 percent, mobile games 39.7 percent and PC Bangs 13.5 percent of market share. Video games, arcade game rooms and PC package games comprised the rest of the market.[34]

Figure 4.3 shows main patterns in the size and growth of the Korean game market during the 2005 to 2016 period. First, the growth of the industry from 2002 to the peak in 2005 was led by growth of online games, which correlated

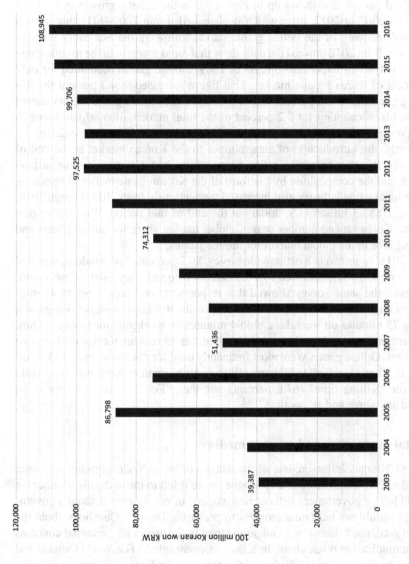

Figure 4.3 Market volume of the Korean game industry, 2003–2016

Source: Korea Game Industry White Papers.

with completion of the Korea Information Infrastructure (KII) project. That project led to substantial increases in the number of homes with broadband internet access and so expanded the online game market far beyond what PC Bangs alone could accommodate. Second, the rather precipitous decrease in the market from 2005–2007 came in large part because of public policy concerns about gambling.

A third pattern that shows up in Figure 4.3 is the healthy growth of the market from 2007 to 2012, after which it plateaued. From 2007–2011 that growth reflected recovery of the online game market following the Sea Story scandal. However, from 2012 onward the data show that rapid increase in the mobile game market played an important role. As of 2009, mobile games accounted for only 4 percent of Korea's game market. That figure increased to 4.3 percent the following year and to 4.8 percent in 2011. In 2012 the mobile game market started to take off, accounting for 8.2 percent of the total market, after which it rose to 23.9 percent in 2013 and 29.2 percent in 2014 and nearly 40 percent in 2016.

Clearly, the introduction of smartphones to the Korean market at the end of 2009 and in other countries a few years earlier had a tremendous impact on growth and the composition by platform of the Korean game market. Exports by the national game industry also increased tremendously from 2011 through 2016, totaling US$3.3 billion U.S. dollars at the end of that period. With that export volume Korea ranked number two in global market share for online games and number four in the global market for mobile games.[35]

By 2015, as shown in a nationwide survey, 86.2 percent of respondents reported using mobile games, followed by 60.3 percent who said they used online games. However, the same survey showed that respondents on average, spent 95 minutes per day on weekdays and 144 minutes on the weekends using online games, versus 75 minutes on weekdays and 94 minutes on weekends for mobile games. Furthermore, respondents indicated clearly different reasons for use of these two platforms. Online games were most frequently used for "Stress release" (75.4 percent) and then "For fun" (64.7 percent) while mobile games were most frequently used for "Killing time" (61.6 percent) and then "For their convenience to be played anywhere and at any time."[36]

Digital journalism and trust in media

In 1787 Thomas Jefferson, one of the authors of the U.S. Constitution who later served as President, wrote to a colleague "were it left to me to decide whether we should have a government without newspapers, or newspapers without a government, I should not hesitate a moment to prefer the latter."[37] Questions about the role of media in a democracy and public trust in the media are perennial concerns of communication researchers. In a famous essay titled "The World Outside and the Pictures in our Heads," Walter Lippmann drew on Plato's allegory of the prisoner in a cave to illustrate how the media portray the world only indirectly.[38] By 2014 the digital revolution led some communication researchers to reframe the issue as "The World Outside and the Pictures in our Networks."[39] The question is more alive than ever in 2018, as political changes around the world, including the election of Donald Trump as president of the U.S. fuel widespread discussion of "fake news."

The Korean case sheds light on several matters that are at the crux of this issue. One is the role of digital media versus more traditional media in informing citizens about affairs in their nation and the world. Another is the question of trust in public institutions, including the media.

Survey data from around the world suggest that younger generations, who grow up using the internet, are more likely to trust it as a source of news and information about the world around them. A ten-country survey in 2006 for the BBC, Reuters and the Media Center showed that more people trust in the media than in their governments, especially in developing countries. Not surprisingly, national television was the most trusted news source overall, trusted by 82 percent of the respondents.[40] Across all ten countries, television was also seen as the most important news source, by 56 percent of the respondents, followed by 21 percent who cited newspapers and only 9 percent who mentioned the internet.

However, the pattern in South Korea was different: 76 percent expressed their highest levels of trust in television, while 64 percent mentioned national and regional newspapers. However, 55 percent also mentioned news websites. Asked which specific news source they consider most trustworthy, South Koreans' responses include KBS television (mentioned by 18 percent), the website NAVER (13 percent), Chosun (10 percent), MBC television (9 percent), DongA and ChoongAng (both 6 percent), DAUM website (5 percent), Hankyoreh (3 percent), South Korea's National TV Station and YTN television (both 3 percent) and Yahoo and the Economist (both 1 percent). South Korea was the only country where websites were so trusted to provide for individuals' news consumption.[41]

In 2007, the Edelman Trust Barometer, a survey of opinion leaders in 18 countries, found that Korea led the world in trusting the internet and blogs in sharing credible information.[42] In 2009, trust in social networks almost doubled from the previous year rising from 24 percent to 45 percent. Globally Edelman found that technology was the most trusted sector, with 76 percent of respondents saying they trusted it in 2009. In the same year, 81 percent of respondents indicated that they trusted the technology sector. This is in keeping with the generally positive portrayal of ICT in most Korean media.[43]

AI versus humans: glimpses of future human-machine interaction

In March of 2016 an event occurred in Seoul that received considerable domestic and worldwide attention, while drawing a stark contrast between human intelligence and the future possibilities presented by artificial intelligence (AI). It was a 5-game Go match (called *Paduk* in Korea) between 18-time world champion Lee Sedol and Alpha Go, a computer program developed by Google's DeepMind machine learning program.[44]

Go, an ancient Chinese board game where two players place black or white playing pieces, called stones, to secure more territory, has more variables than chess, and until the match between Lee Sedol and Alpha go was considered more difficult to teach to computers. The news that artificial intelligence had defeated humanity's best Go player sent shock waves throughout the world, and most especially throughout Korea.[45] The shock in Korea was deeply cultural and historic in nature. In this country, the game features television contracts and corporate sponsors. Myongji University even has a Department of Paduk Studies, founded in 1997, which enrolls students from Korea and around the world. When South Korea's top player was bested by a computer's AI, it shook thousands of years of traditional cultural understandings, not only in Korea, but around the Asian region and the world.[46]

In describing the match, *Wired* headlined its article claiming that "In two moves, AlphaGo and Lee Sedol redefined the future." The Google machine won four out of the five games. In game two it made a move that no human ever would, perfectly demonstrating the enormous power and mysterious talents of modern AI. However, in game four, Lee Sedol made a move that no machine would ever expect.[47]

The *Wired* article goes on to note that the machine's victory in the best-of-five series marked the first time that a machine had beaten the very best at this ancient and complex game. It surprised many experts who thought such a defeat would not happen for another decade. Also, it did so using technologies that are already changing Google, Facebook and Microsoft, but are poised to reinvent everything from robotics to scientific research. In a concluding comment, the article notes that "This isn't human versus machine. It's human *and* machine."[48]

Shortly after the shock of Alpha Go's victory over Lee Sedol, President Park Geun Hye assembled senior government officials and leaders of the technology industry to announce plans for investing 3.5 trillion won (about US$3 billion) in AI research and development over the next five years. The President told reporters at the meeting, "Above all, Korean society is ironically lucky, that thanks to the "AlphaGo shock" we have learned the importance of AI before it is too late." She added that the game between AlphaGo and Lee Sedol was a "watershed moment" in the imminent fourth industrial revolution.[49]

Another event with implications for the future of AI and its implications for humans and society took place over a year later in November of 2017. Sejong University in Seoul hosted a match between AI systems and professional StarCraft player Song Byung-gu. Song defeated four different bots in the first such match to put AI systems against a human. The AI research community considered StarCraft a particularly difficult game for bots too master.

> Unlike Go, which allows bots and human players to see the main board and devote time to formulating a strategy, StarCraft requires players to use their memory, devise their strategy, and plan ahead simultaneously, all inside a constrained, simulated world. As a result, researchers view StarCraft as an efficient tool to help AI advance.[50]

As of 2017 many observers thought it would take five to ten years for an AI-driven bot to defeat the best human players of StarCraft. However, in January 2019

DeepMind, the company that had developed AlphaGo, revealed that it had created an AI called AlphaStar that could play StarCraft II well enough to beat some of the best human players in the world. Progress in AI had progressed faster than most observers thought possible.[51]

How language shapes media use

Language as a key element of culture exerts a strong influence on how the internet and other media are used. As noted in the ITU's *Broadband Korea* case study

> The top 10 web sites accessed by Korean users are all in Korean. The number of domains registered using. KR – almost exclusively in the Korean language – ranks the nation fifth in the world. Not only has this driven use, but it has also reduced the need for expensive international circuits. It also suggests that in many ways the Internet in Korea is actually one big Intranet with most users preferring to access local sites.[52]

Korean netizens' strong preference for Korean language web content largely explains why the web portal Naver, rather than Google, dominates the search market here. This preference persists despite the fact that Naver offers almost exclusively Korean language content and does not actually "search" the internet. Korea is one of only four national markets in the world today where Google's internet search does not have a strong share of the market. The others in what the *Financial Times* termed the "non-Google World," are China, Russia and the Czech Republic.[53] In each of those countries, local companies designed technology to work in the local language.

A second reason for the tremendous popularity of Naver in South Korea is that it caters to the homogeneous and group-oriented culture. Its most popular feature is called "Knowledge-in." Using this feature, a user can submit a question and instantly receive a reply from the database that benefits from the collective intelligence of millions of other Korean users. It does an outstanding job of telling a Korean user what other Koreans are thinking on any given topic.

One harsh critique of the impact of broadband in Korea suggested that it has actually narrowed the Korean mindset. In this view the internet, instead of opening up Korea and Koreans, has made them more tribal and less global than before the internet. Huer argues that the protective anonymity of the internet has given Korean netizens a perfect sanctum from which to fire their fury and vengefulness that have long been denied.[54]

Finally, it is worth noting that many Korean web surfers became comfortable with Naver before Google entered the Korean market or they became aware of it on their own. Korea's lead over the U.S. and other nations in building high-speed broadband networks was instrumental in creating this situation. Nevertheless, the continued success of Naver in the South Korean internet market underscores the crucial role of language and culture in shaping media use.

Commerce, society and politics in the mobile revolution

South Korea's consumer culture started changing with completion of the nation-wide, digitally switched PSTN in 1987, the start of a credit card verification service called EasyCheck (now the Korea Information and Communications Company, Ltd.)[55] and the arrival of credit cards. But it has come a long, long way over the years.

By 2014 Korea ranked number one in the world in number of credit card payments per person. Koreans on average made 147 payments per year with their credit cards, the most among 18 countries. Canada and the U.S. followed with 89.9 and 83.5 respectively. Korea also ranked number three in the amount of payments made by credit card at US$8,625.[56]

An alternative to credit cards, the electronic wallet, made significant inroads into South Korea's consumer culture after the turn of the millennium. Korea's mobile operators offer a service called the "Cellular Small Payment Service." It allows subscribers to pay for online goods using authorization codes sent by SMS. This service was particularly popular with younger Koreans, who could add such costs to their mobile telephone bills without needing a credit card. As of 2006, some 23 million Koreans or nearly half the population was using one of five competing cell-phone systems to make payments ranging from a few cents to US$120.[57]

Mobile payments increased dramatically after the introduction in 2015 of services like Samsung Pay, which allow users to pay for goods or services at the point of sale. Such services allow users to register credit or debit cards on their smartphones. Payments can then be made by placing their smartphones near a card reader. Samsung Pay uses both near field communication and magnetic secure transmission technology to transmit payment information between the phone and payment terminal. As of March 2018, Samsung Pay had secured more than 10 million users.[58]

It is deeply ironic that South Korea led the world in moving toward online banking in the late 1990s, based on its world leading broadband internet infrastructure. Unfortunately it adopted Microsoft's ActiveX software framework. Although ActiveX was cutting edge in 1996, the security of that software was so flawed that Microsoft itself removed it from its next generation browser and warned users around the world against using it. To make matters even worse, the use of ActiveX required the use of Internet Explorer, just as users all around the world were moving toward faster and more modern browsers like Chrome and Firefox. By the early years of the 21st century, Korea found itself with an online banking system that relied on obsolete technology. A 2016 article in *Forbes* described the process as follows:

> South Korea's online banking infrastructure is a mutation of its own. If a user wants to buy socks, wire money or pay the gas bill, they must pull out a slew of identity checks, from a simple username and password to online digital certificates, unique number cards and one-time password devices that spit

out a six-digit code for a few seconds. There might even be a text message or phone call confirmation involved, and all this can only be done after down-loading a mountain of cybersecurity programs – limited to your designated device – that some experts say do more harm than good.[59]

It was only in 2015 that Korean regulators eliminated the requirement for finan-cial institutions to use ActiveX. That same year the nation's Financial Supervisory Commission revised its regulations to allow the formation of internet banks. Near year's end, Kakao was given preliminary approval to launch the internet-only Kakao Bank. The intent was to open up Korea's FinTech (financial technology services) sector to innovation.[60]

Thanks in part to the earlier completion of high-speed broadband networks, South Korea pioneered in the field of social networking, as this phenomenon swept through the nation's internet about four years before it reached the United States. Nothing illustrated South Korean's comfort with cyberspace and its new media environment better than Cyworld, launched in 1999.

As of late 2008 nearly half of the country's population, 90 percent of Koreans in their 20s, and many socialites or celebrities used Cyworld. Young people, when meeting for the first time would frequently ask for another's "cyaddress" rather than their phone number. Some 40,000 companies, non-profit organizations, gov-ernment agencies and universities had joined Cyworld to promote their businesses and activities.[61]

Unlike its Western social networking equivalents, Cyworld had cute avatars and "mini-rooms," known as *mini-homepi*, that are interconnected with other friends' and family pages. Friends can visit other friends' mini-rooms, which are places in cyberspace that often reflect offline spaces. As one researcher noted, although Cyworld's mini-rooms are filled with cute customizations that might be seen as childish in a Western context, in Korea the cute content is consumed by both young and old users.[62] Koreans seem to associate the cuteness with part of "a struggle to humanize and socialize technological spaces, to highlight the mediated role of intimacy regardless of technological interference."[63]

In August of 2006 Cyworld officially launched a version of its site in the United States. However, that effort at market entry had failed by the fall of 2008.[64] This was another powerful illustration of the strong role that culture and language play in the shaping of cyberspace.

By January of 2012, Facebook passed Cyworld in popularity among Korean PC users. Figure 4.4 is based on Nielsen panel data, and it also shows that, by December of 2013, 80 percent of PC users in the panel used Facebook, compared with only 20 percent who continued to use Cyworld.

In addition to internet addiction, Korea has had its share of experience with cybercrime. In early 2009, there were press reports of increasing activities by cyber-gangs, based in China, affecting Korean internet businesses. Servers receive a Distributed Denial of Service (DDos) attack, followed by a blackmail message that asks for a certain sum of money to stop the attack. For example, the owner of a flower delivery service named K received a blackmail message. He disregarded

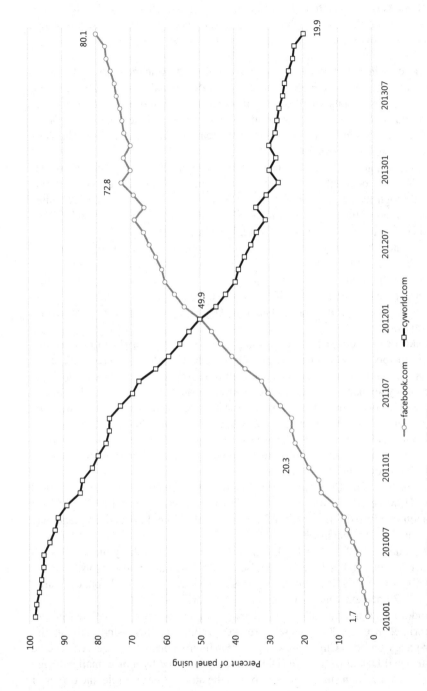

Figure 4.4 Trends in PC use of Facebook and Cyworld in Korea

Source: Nielsen Korea.

it, and his server was hit with a DDos attack. He was able to avoid further attack by transmitting 3 million won to the attacker. Industry insiders say that these types of attacks began in 2008 and that there were ten or more occurring each month.[65] In 2007 the government reported that 1,823 or 1.1 percent of organizations surveyed experienced damage from denial of service attacks.[66]

"Malware" is a general term for software inserted into an information system that can cause harm to that system or other systems or can subvert them for use other than that intended by their owners. Different types of malware are described as worms, viruses, Trojan horses, backdoors, keystroke loggers, rootkits and spyware.[67] Studies by Google and other organizations lead to the conclusion that about 80 percent of all web-based malware is being hosted on innocent but compromised websites, unbeknownst to their owners.[68]

Virus infections and various forms of worms and malware are a continuing problem in South Korea. One of the main reasons is that there are still a large number of computers with internet access that are not using proper anti-virus, malware and firewall protection.

In 2007, a comprehensive survey showed that 16 percent of establishments in South Korea had experienced damage from an attack by a computer virus, worm or Trojan horse. This amounted to approximately 36,000 establishments.[69]

In an effort to reduce damage from botnets, two measures were undertaken in Korea. First, to reduce damage from vulnerabilities in the widely used Windows operating system, the Korea Internet Security Center (KrCERT/CC) and Microsoft Korea collaborated to develop and deploy the Automated Security Update Program (ASUP) to home and small business users. When users visit major Korean websites, such as portals or online game sites, a pop-up window appears on the screen to confirm installation of the ASUP. A user only needs to click once, without modifying the Windows automatic update settings.

The second measure was adoption of the sinkhole system. It works to prevent botnets from connecting to botnet command and control servers by subverting the IP address of such command and control servers. The botnet infection rate in Korea dropped considerably following adoption of the sinkhole system in 2005.

A third countermeasure being used in Korea is the use of MC Finder, which detects malware on compromised websites. As of 2008, MC Finder detected malware on an average of 500 websites per month. The Korea Internet Security Center was sharing such malware patterns with Google and three major Korean web portals.[70]

Despite all the efforts described above, on July 4, 2009, a wave of cyber-attacks was unleashed on 27 American and South Korean government agencies and commercial websites. In South Korea, at least 11 major sites slowed or crashed, including the presidential Blue House, the Defense Ministry, the National Assembly, Shinhan Bank, the mass circulation newspaper *Chosun Ilbo* and the internet portal Naver, according to the government's Korea Information Security Agency. In the United States, the websites of the Treasury Department, Secret Service, Federal Trade Commission and Transportation Department were all affected, and the White House was also included in the attacks.

South Korea's National Intelligence Service released a statement saying that "This is not a simple attack by an individual hacker, but appears to be thoroughly planned and executed by a specific organization or on a state level," adding that it was cooperating with the American authorities to investigate the attacks.[71]

Months later, in October of 2009, the head of Korea's NIS testified before the National Assembly that in tracking down the routes of the July DDOS attacks, his agency had found a route coming from China. This turned out to be an IP that North Korea's Ministry of Posts and Telecommunications was renting.[72]

As in other countries where internet usage is at high levels, Korea faces a continuing problem with spam, both to regular e-mail accounts and in the form of calls to mobile phones. In recent years, Korea has appeared on the list of top spam-relaying countries. For example, the security firm Sophos, the United States ranked first in the world, relaying 13.1 percent of global spam, followed in order by India, which relayed 7.3 percent, Brazil 6.8 percent and South Korea 4.8 percent. Although these four countries led the world in relaying of spam, they accounted for only a little more than 30 percent of global spam.[73]

Korea's evolving information culture also confronts the challenge posed by cyber-bullying. The main distinction South Korea can claim, as with the malady of internet addiction, is that it began to experience cyber-bullying about four years before the United States did. That is the approximate time lag between the introduction of Cyworld in Korea and the corresponding start of Myspace or Facebook in the U.S.

In order to deal with the problems of cyber-bullying and misinformation on the internet, the Korea Communications Commission in the fall of 2008 mandated that all internet sites with more than 100,000 visitors impose real-name registrations for their message boards and chat rooms beginning in April of 2009. On that date, an amendment to South Korea's Act on the Promotion of Information and Communication Network Utilization and User Protection went into effect,[74] and Korea became the first country in the world to implement a "real-name" system under which any South Korean can post comments only after they enter their national registration number. One example of this policy was that Korea became the first country in the world where Google requires real-name identification for its YouTube Korea site. Prior to April 1, 2009, users of Google Korea were able to post materials simply by giving an ID, a password and an e-mail address. Google's head office reportedly explored various means of bypassing the "real-name registration system," arguing that freedom of expression should be experienced globally by all users. Google even at one point suggested shutting down YouTube services in South Korea. The country director of Google Korea said that "Google respects users' rights and freedom of expression to the fullest, and at the same time it also respects local regulations."[75]

Building an information society is a human endeavor and at its very heart involves language, culture and communication. The topics broached in this chapter reinforce that point along with the constant interplay of human culture and technology. One finding seems quite certain. That is the strength and resilience of the Korean culture in the face of rapid and sweeping technological change and development.

Beyond resilience, the manner in which Korea's language and culture are shaping its information society should be a profoundly hopeful sign to those in other nations around the world. It suggests that the future global information society may indeed embrace the diversity and richness of human languages and culture.

Notes

1 Dobbs, Richard, James Manyika and Jonathan Woetzel. (2016, April). *The four global forces breaking all the trends*. McKinsey and Company. Retrieved December 23, 2016, from www.mckinsey.com/business-functions/strategy-and-corporate-finance/ our-insights/the-four-global-forces-breaking-all-the-trends

2 Song, N. A. (1994). The nature of urbanization in South Korea. In A. K. Dutt, *The Asian City: Processes of development, characteristics and planning*. Dordrecht, The Netherlands: Kluwer Academic Publishers.

3 Castells, M. (2007). Space of flows, space of places: Materials for a theory of urbanism in the information age. In W. W. Braham, *Rethinking technology: A reader in architectural theory*. London: Routledge, pp. 440–456.

4 Kang, M. G. (1998, June). Understanding urban problems in Korea: Continuity and change. *Development and Society*, 27(1), p. 100.

5 Wikipedia. (n.d.). *Interstate highway system*. Retrieved December 30, 2016, from Wikipedia, the free encyclopedia: https://en.wikipedia.org/wiki/Interstate_Highway_System

6 Gore, A. (1994, Jan 11). *Remarks as delivered by Vice President Al Gore to the Superhighway Summit Royce Hall, UCLA Los Angeles California*. Retrieved December 31, 2016, from clinton1.nara.gov: https://clinton1.nara.gov/White_House/EOP/OVP/other/ superhig.html

7 Henderson, J. V. (2012). Measuring economic growth from outer space. *American Economic Review*, pp. 994–1028.

8 Do, M. M. (2016, August 13). *Coverage maps for wired/wireless service now available from South Korea's big 3 operators*. Retrieved December 31, 2016, from Netmanias: www.netmanias.com/en/?m=view&id=blog&no=10326

9 OpenSignal. (2016, February). *The state of LTE (February 2016)*. Retrieved October 23, 2016, from OpenSignal: https://opensignal.com/reports/2016/02/state-of-lte-q4-2015/

10 Hwang, J. S. (2013). *Smart cities Seoul: A case study*. Geneva: ITU.

11 Mitchell, S. N. W. (2013). *The internet of everything for cities*. San Jose, CA: Cisco. Retrieved June 6, 2019, from https://www.cisco.com/c/dam/en_us/solutions/industries/ docs/gov/everything-for-cities.pdf

12 Hwang, J. S. (2013). *Smart cities Seoul: A case study*. Geneva: ITU, p. 1.

13 *U-City History*. (n.d.). Retrieved December 31, 2016, from Korea Smart City Association: www.ucta.or.kr/en/introduce/history.php

14 U-KOREA Master Plan to Achieve the World's First Ubiquitous Society. (2006, May). Ministry of Information and Communications, Republic of Korea.

15 U-KOREA Master Plan to Achieve the World's First Ubiquitous Society. (2006, May). Ministry of Information and Communications, Republic of Korea, p. 3.

16 U-KOREA Master Plan to Achieve the World's First Ubiquitous Society. (2006, May). Ministry of Information and Communications, Republic of Korea, p. 9.

17 U-KOREA Master Plan to Achieve the World's First Ubiquitous Society. (2006, May). Ministry of Information and Communications, Republic of Korea, p. 10.

18 Korean ministries team up to build u-cities nationwide. *Korea IT Times*, March 31, 2006. Retrieved from www.koreaittimes.com/story/2472/korean-ministries-team-build-u-cities-nationwide

19 Chung, Myung-je. Leading global u-city. *Korea IT Times*, August 7, 2009. Retrieved from www.koreaittimes.com/story/4371/leading-global-u-city

20 Chung, Myung-je. Leading global u-city. *Korea IT Times*, August 7, 2009. Retrieved from www.koreaittimes.com/story/4371/leading-global-u-city

21 2008–09 Accomplishments, 2008–2013 Five Year Plan, Stonybrook, State University of New York, May 2009, p. 27.

22 Hwang, J. S. (2013). *Smart cities Seoul: A case study*. Geneva: ITU, p. 2.

23 Hwang, J. S. (2013). *Smart cities Seoul: A case study*. Geneva: ITU, p. 6.

24 Limb, Jae-un. (2009, November 20). From digital screens to sculpture gardens, art the masses can enjoy. *Joongang Daily*.

25 Interview with Han, Hoon. Visiting Professor, Seoul National University Graduate School of Engineering Practice by James F. Larson, January 4, 2017.

26 Rea, S. C. (2016). Crafting stars: South Korean e-sports and the emergence of a digital gaming culture. *Education about Asia*, 21(2), pp. 22–27.

27 Huhh, Jun-Sok. (2008, January). Culture and business of PC bangs in Korea. *Games and Culture*, 3(1), p. 28.

28 Rea, S. C. (2016). Crafting stars: South Korean e-sports and the emergence of a digital gaming culture. *Education about Asia*, 21(2), p. 22.

29 Chee, Florence. (2006). The games we plan online and offline: Making *Wang-tta* in Korea. *Popular Communication*, 4(3), p. 231.

30 Rea, S. C. (2016). Crafting stars: South Korean e-sports and the emergence of a digital gaming culture. *Education about Asia*, 21(2), p. 23.

31 Jang, D. S. (2011, September 3). At one school, students are developers. *Korea Joongang Daily*.

32 Rea, S. C. (2016). Crafting stars: South Korean e-sports and the emergence of a digital gaming culture. *Education about Asia*, 21(2), p. 24.

33 Sohn, Ji Young. (2018, June 20). Korean market grows for online, mobile games. *Korea Herald*.

34 2017 White Paper on Korean Games (Summary), Korea Creative Content Agency (KOCCA), 2017, p. 6.

35 2017 White Paper on Korean Games (Summary), Korea Creative Content Agency (KOCCA), 2017, pp. 8–10.

36 Korea Creative Content Agency. (2015). *2015 white paper on Korean games*. Naju, Republic of Korea: Korea Creative Content Agency, p. 26.

37 Jefferson, T. (n.d.). *Amendment I (speech and press) document 8 Thomas Jefferson to Edward Carrington 16 Jan 1787*. Retrieved January 6, 2017, from The Founders Constitution: http://press-pubs.uchicago.edu/founders/documents/amendI_speechs8.html

38 Lippmann, W. (1922). *Public opinion*. New York: Harcourt, Brace, chapter 1.

39 Turner, Fred. The world outside and the pictures in our networks. Chapter 13 In Tarleton Gillespie, Pablo J. Boczkowski and Kirsten A. Foot, *Media technologies: Essays on communication, materiality and society*. Cambridge, MA: MIT Press, pp. 251–260.

40 BBC, Reuters. (2006, May 3). *Media center poll: Trust in the media*. Retrieved from www.globescan.com/news_archives/bbcreut.html

41 BBC, Reuters. (2006, May 3). *Media center poll: Trust in the media*. Retrieved from www.globescan.com/news_archives/bbcreut_country.html

42 Edelman. 2009 Edelman Trust barometer: Korea report. Retrieved from www.slide share.net/Edelmankorea/2009-the-edelman-trust-barometer-korea-report

43 Edelman. 2009 Edelman Trust barometer: Korea Report. Retrieved from www.slide share.net/Edelmankorea/2009-the-edelman-trust-barometer-korea-report

44 Wikipedia. AlphaGo versus Lee Sedol. Retrieved on October 27, 2018, from https://en.wikipedia.org/wiki/AlphaGo_versus_Lee_Sedol

45 Zastrow, M. (2016, March 9). I'm in shock! How an AI beat the world's best human at Go. *New Scientist*.

46 Zastrow, M. (2016, March 15). How victory for Google's Go AI is stoking fear in South Korea. *New Scientist*.

47 Intwomoves,AlphaGoandLeeSedolredefinedthefuture.*Wired*,March16,2016.Retrieved from www.wired.com/2016/03/two-moves-alphago-lee-sedol-redefined-future/

48 In two moves, AlphaGo and Lee Sedol redefined the future. *Wired*, March 16, 2016. Retrieved from www.wired.com/2016/03/two-moves-alphago-lee-sedol-redefined-future/

49 Iglauer, P. (2016, March 22). *South Korea promises $3b for AI R&D after AlphaGo 'shock'*. Retrieved February 10, 2017, from ZDNet: www.zdnet.com/article/south-korea-promises-3b-for-ai-r-d-after-alphago-shock/

50 Kim, Yoochul and Minhyung Lee. (2017, November 1). Humans are still better than AI at StarCraft – for now. *MIT Technology Review*. Retrieved October 27, 2018, from www.technologyreview.com/s/609242/humans-are-still-better-than-ai-at-starcraftfor-now/

51 Dickson, Ben. (2019, January 29). AI defeated humans at Starcraft II. Here's why it matters. *TechTalks*. Retrieved from https://bdtechtalks.com/2019/01/28/deepmind-alphastar-ai-starcraft-2/

52 *Broadband Korea: Internet case study*. ITU, March 2003, p. 11.

53 Google still struggling to conquer outposts. *The Financial Times*, September 16, 2008. Retrieved from www.ft.com/content/99d3e98a-8406-11dd-bf00-000077b07658

54 Huer, Jon. Has internet closed Korea more? *Korea Times*, April 26, 2009. Retrieved from www.koreatimes.co.kr/www/news/opinon/2009/04/272_43854.html

55 Korea Information and Communications Company. EasyCheck (Korean language website). Retrieved from www.kicc.co.kr/eng/company_history.jsp

56 Kim, H. S. (2014, May 12). Koreans, world's number one in number of credit card payments per person. *The Kyunghyang Shinmun*.

57 Moon, Ihlwan. (2006, March 2). In Korea, cell phones get a new charge. *Business Week*.

58 Sohn, Ji-young. (2018, April 3). Korea's mobile payments market grows in both offline, online. *Korea Herald*.

59 Ramirez, Elaine. (2016, November 30). South Korea's online banking system is stuck in 1996. *Forbes*. Retrieved from www.forbes.com/sites/elaineramirez/2016/11/30/south-koreas-online-banking-system-is-stuck-in-1996/#2d3d6000527c

60 Iglauer, Philip. (2015, December 2). Kakao to launch web-based bank. *ZDNet*. Retrieved from www.zdnet.com/article/kakao-to-launch-web-based-bank/

61 Hall, Kenji, Moon Ihlwan and Bruce Einhorn. (2006, September 11). In Asia, Myspace clones stalk cyberspace. *Business Week*. Retrieved from www.businessweek.com/technology/content/sep2006/tc20060911_808191.htm

62 Hjorth, Lisa. The game of being mobile: One media history of gaming and mobile technologies in Asia-Pacific. *Convergence: The International Journal of Research into New Media Technologies*, p. 374. Retrieved from http://con.sagepub.com/cgi/content/abstract/13/4/369

63 Hjorth, Lisa. The game of being mobile: One media history of gaming and mobile technologies in Asia-Pacific. *Convergence: The International Journal of Research into New Media Technologies*, p. 375. Retrieved from http://con.sagepub.com/cgi/content/abstract/13/4/369

64 Malik, Om. Cyworld packs up from U.S., retreats to Korea. Retrieved from http://gigaom.com/2008/11/09/cyworld-packs-up-from-us-retreats-to-korea/. Date of the Gigaom post is November 9, 2008

65 Jang, Dong-joon and Kim In Soon. (2009, February 18). Cyber gangs to accelerate. *Korea IT News*. Retrieved from http://english.etnews.co.kr/news/detail.html?id=200902180006

66 2008 Yearbook of Information Society Statistics, National Information Society Agency, part 6.

67 Computer Viruses and other Malicious Software: A Threat to the Internet Economy, OECD Report, March 2009, p. 21.

68 Computer Viruses and other Malicious Software: A Threat to the Internet Economy, OECD Report, March 2009, p. 24.
69 2008 Yearbook of Information Society Statistics, National Information Society Agency, part 6.
70 Computer Viruses and other Malicious Software: A Threat to the Internet Economy, OECD Report, March 2009, p. 175.
71 Choe, Sang-Hun. (2009, July 8). Cyberattacks Jam government and commercial web sites in U.S. and South Korea. *New York Times*. Retrieved from www.nytimes.com/2009/07/09/technology/09cyber.html?scp=1&sq=July%204%20cyber%20attacks&st=cse
72 Kim, Sue-Young. (2009, October 30). Spy chief says cyber attacks work of North Korea. *The Korea Times*. Retrieved from www.koreatimes.co.kr/www/news/nation/2009/10/113_54596.html
73 Sophos. Dirty Dozen. Retrieved July 13, 2010, from www.sophos.com/pressoffice/news/articles/2010/04/dirty-dozen.html
74 Hankyoreh. Google compromises on Internet free speech in S. Korea. Retrieved from http://english.hani.co.kr/arti/english_edition/e_international/346930.html
75 Google compromises on internet free speech in South Korea, *The Hankyoreh* English web edition, March 31, 2009. Retrieved from http://english.hani.co.kr/arti/english_edition/e_international/346930.html

5 Education, research and development for the hyperconnected era

Education is the basic process of the information age and is central to an understanding of Korea's remarkable ICT-led transformation. The exponential increases in the human ability to store, compute and communicate digital information that arose after the mid-20th century profoundly changed the practices and possibilities in education all around the world. The new digital networks expanded the geographical scope and accessibility of education and effectively brought the walls of brick and mortar "ivory tower" universities crashing down.

As the Korean case powerfully illustrates, education provides the building blocks for the information society. More specifically, as noted by a World Bank/OECD study, education helps form the following four key pillars of the knowledge economy:

- An economic and institutional regime with incentives for the use of existing knowledge and the creation of new knowledge.
- Education, training and human resource management (an educated, entrepreneurial population that can both use existing knowledge and create new knowledge).
- A dynamic information infrastructure, to facilitate effective communication, dissemination and processing of information.
- An efficient innovation system, comprising firms, science and research centers, universities, think tanks, consultants and other organizations.[1]

From the rubble of the Korean War and half a century of Japanese occupation, South Korea faced the task of re-building its entire system of schools. Lacking natural resources, South Korea turned to education and knowledge as the key engine of economic growth.[2]

This chapter is organized around four main topics. First, it tells the story of how Korea built up its formal educational system following the Korean War. Second, it addresses the revolutionary changes that information and communication technologies bring to education globally and in Korea. Third, the chapter discusses the critical role that education plays in building the ICT sector itself. Finally, it examines the new central role of research and development in the digitally networked era.

Building the formal educational system

When the armistice that ended hostilities in the Korean War was signed on July 27, 1953, South Korea faced an extraordinary challenge simply to build an educational system. There were two major reasons for this situation. One was the half-century of Japanese colonial rule in Korea, which ended with Japan's defeat in World War II. The other was the widespread destruction and dislocation of families caused by the Korean War.

The legacy of colonial rule and the Korean War

Consider the state of Korean education at the end of World War II in 1945, which ended Japan's colonial rule. Only 64.0 percent of elementary school-aged children were enrolled in school. This percentage plummeted to 3.2 percent for secondary education and 0.18 percent for higher education.[3] One survey estimated that the population of South Korea aged 13 years old and above was 15 million and that 53 percent of that population (8 million) was illiterate. Those who had a secondary education or more accounted for only 12.6 percent of the population.[4]

Faced with this desperate situation, the U.S Military Government and the Republic of Korea set the expansion of elementary education as their number one task for educational development. This was especially challenging as 68 percent of the school buildings had been destroyed during the Korean War.[5] During the war itself, displaced families had set up schools in tents and made an effort to continue their children's schooling. However, many children were unable to attend school during the conflict itself and thus graduated from high school two years late.

Today's universal literacy and high levels of educational achievement in South Korea need to be measured against the legacy of colonial education policy and the devastation of the Korean War. Defining universal enrollment as attaining a 90 percent enrollment rate, Korea achieved universal elementary education by 1957, universal middle school education by 1990 and universal high school education by 1999. It is no exaggeration to say that Korea's first move in modernization came in education.[6]

As the nation began to develop, the requirements of industry and government for technically trained people increased, and so did the demand for university-level education. Accordingly, enrollment in progressively higher levels of education increased as South Korea moved from an emphasis on heavy industry in the 1970s to electronics and information industries beginning in the 1980s and through the 1990s.

By 2007 Korea ranked fourth among OECD countries, behind the Russian Federation, Canada and Japan with all four countries having more than 50 percent of the population attain a tertiary degree. Along with Japan, France and Ireland Korea showed a gap of more than 25 percent between the older and younger age groupings, indicating the rapid expansion of tertiary education in recent years.[7]

Korea's efforts to produce skilled information workers

Strengthening vocational education became a priority in the economic development plans of the 1970s and vocational junior colleges were set up during that decade to supply technicians for the Heavy Chemical Industry (HCI) drive. That plan explicitly acknowledged the need to upgrade technology and the technical workforce. The framework of vocational education was institutionalized in 1976 through enactment of the Basic Vocational Training Act, which was wholly amended in 1981.[8]

The increased need to provide students with scientific, technological and other skills for the information age is a common denominator among the 30 member nations in the OECD. In a particularly telling statistic, South Korea led the world in 2006 in the proportion of tertiary science graduates in the 25–34 age group per 100,000 employed people in the same age cohort. This measure is one way that the OECD gauges the output high-level skills by different educational systems.[9]

Part of Korea's approach to education for the information age has been to change the focus and curricula of its schools at all levels to meet the changing human resource requirements. At the high school level, Korea has general high schools, vocational high schools, science high schools and other specialized high schools. Courses offered at the vocational high schools were diversified over recent decades to include information technology, robotics, animation, films, cooking, beauty, tourism, horse care and so forth to meet the demands of a rapidly changing industrial society.

Junior colleges, which have mostly two-year and some three-year post-secondary programs, are a direct outgrowth of the increased demand for technical manpower with rapid industrialization. As of 2016 there were 138 junior colleges in Korea, a majority of which were private institutions.[10]

A large focus of junior college education was on industry-academia cooperative efforts. These cooperative programs between junior colleges and industries include student internships, industrial field training for junior college faculty, education of industry employees at junior colleges, joint research, exchanges of technology and information and suggestions for the curriculum by industries. Over the years, 80 percent or more of junior college graduates gained employment after graduation, a higher percentage than for four-year colleges and universities.[11]

The Ministry of Education administers public universities in South Korea, but the Ministry of Science and Technology over the years has contributed funds to university science and technology programs, both public and private, through the Korean Science and Engineering Foundation (KOSEF), Korea's equivalent of the U.S. National Science Foundation. As of 1997, KOSEF had established approximately 30 university science and technology centers of excellence, which it funded annually at the million-dollar level. Those centers were required to collaborate with at least three other institutions and were strongly encouraged to attract supplemental support from industry.[12] Centers with the best reputations include the Korean Advanced Institute of Science and Technology (KAIST) and Seoul National University (SNU).

South Korea's universities were generally modeled after those in the United States. Like their American counterparts, major private and public universities in Korea have played a major role in basic research and development across virtually all of the major fields of scientific and technological research. University faculty and leading university research institutes in South Korea have strong ties with all of the major governmental and private research institutes. One factor that has strengthened such ties both domestically and globally is the large number of U.S.-trained Ph.D.s in the nation's universities and research institutes.

ICT in education: the digital transformation of learning

Evidence of the impact of the digital network revolution on education abounds, both globally and in South Korea. To help place Korea's experience with ICT in education in context, this section of the chapter looks first at some of the important global changes and then traces the influence of ICT on education in Korea.

The global transformation

The driving force in the transformation of education today is the continuing transition from an industrial mass media economy to the networked, knowledge-based sharing economy. Universities and schools at all levels can no longer exist only as brick and mortar campuses, isolated from the rest of the world. Instead, the new model for university education places the institution in a Silicon Valley-type environment that involves active collaboration with private corporations, governments and non-governmental organizations.

In a 2005 report, *The Economist* noted four reasons for the sea change taking place in universities. One is the democratization of education as epitomized by the increasing proportion of adults with higher education degrees and the spread of massive open online courses (MOOCs). A second is the rise of the knowledge economy, sometimes called the "soft revolution," in which knowledge replaces physical resources as the main driver of economic growth. The third factor is globalization in which ease of communication via the internet and improvements in travel contributed to large increases in the number of students studying abroad. The fourth and final reason for the revolution in higher education is competition, in which private companies compete for both research grants and students.[13]

A McKinsey report, after referring to the huge potential of digital tools to transform learning, noted that digital learning tools themselves are not that new. "What is new – and disruptively so – is the fact that the content of learning is moving to the cloud, becoming accessible across multiple devices and teaching environments and often being generated, shared, and continually updated by users themselves."[14] Both industry reports and commentary from school and university administrators document the dramatic shifts, even upheaval in the education publishing industry[15] and in the publication of academic research generally.

Stanford University offers one important model of the influence of ICT on education. It was not by accident that Silicon Valley, the world's model for start-up ventures and university-industry collaboration, originated and grew adjacent to Stanford. Edward Fiske, longtime education editor of the *New York Times* when asked about the differences between Harvard and the other revered East Coast Ivy League schools and Stanford University, replied, "I think the point that I make about Stanford is that it is the first great American university."[16]

Elaborating on Fiske's point, John Hennessy, Stanford's President from 2000–2016 noted that, until the mid-20th century, Stanford was considered a good regional school, but not a world-class center for innovative research and learning. He went on to suggest that "An institution must also make wise choices. Under the leadership of engineering dean and provost Frederick Terman, science and engineering were emphasized and became a key to the development not only of Stanford, but the entire region."[17] In addition to the emphasis on science and engineering, Terman also encouraged the development and pursuit of government funding and support of scientific research. The Silicon Valley model of university-corporate-government collaboration has been replicated many times over, not only with research universities in the United States, but also around the world. As discussed in Chapter 2, Stanford's Frederick Terman himself played a key role in the international report leading to the foundation of KAIST.

Education at global scale: lifelong learning, MOOCs and education hubs

A key feature of the digital network revolution is the growing importance of digital platforms in a growing range of markets and sectors. The platform revolution places information and knowledge at the center of the new economic and social order. A platform, simply defined, is a business that enables value-creating interactions between external producers and consumers.[18] Through the 20th century industrial era, businesses employed a linear pipeline model in which a firm initially designed a product or service, then put it up for sale or devised a system to deliver the service, after which customers showed up and made purchases. Today, platform structures in which producers, consumers and the platform itself interact have replaced the pipeline in many industries. Prominent examples of platform-based businesses include Google, Apple, Amazon and Airbnb.[19]

Education is the prime example of a large industry ripe for platform disruption. The drive to build education platforms such as Skillshare, Udemy, Coursera, EdX and Khan Academy is well under way, with varying results. Furthermore, the platform-based unbundling of educational activities is separating the learning of specific skills from reliance on traditional universities, enabling a larger number of citizens to access learning opportunities more flexibly in a wider range of contexts.[20]

Existing education systems have not kept pace with such changes. Neither possession of a college degree nor vocational training are able to meet the growing demand for lifelong learning to gain the skills and knowledge needed to work with continually changing technologies. To remain competitive and provide workers the best chance for success at all skill levels, nations need to offer training and career-focused education throughout the peoples' working lives.

The internet, by allowing increasingly realistic and immersive video conferencing, effectively changes our perception of distance. Given this inherently global scope of the internet, the arrival of MOOCs was probably inevitable. Open online courseware makes it possible for anyone with a sufficiently fast connection to the internet to take a broad range of courses. For skills-oriented courses such as computer programming, the completion of an online course may frequently carry just as much weight on a student's resume and have just as much impact on their employability as a course taken in a recognized and accredited college or university.

Global access to university-level course material has existed for some time. For example, MIT OpenCourseWare published its first 50 courses in 2002.[21] The origins of the MOOC revolution are frequently dated from 2011 when Stanford University Professor Sebastian Thrun offered a free course on artificial intelligence (AI) online. It was similar to the course he taught at Stanford where about 200 students enrolled. When the AI's MOOC started, there were 160,000 students from virtually every country in the world enrolled. Based on this experience and the irony that Stanford students were paying US$50,000 or more annually to attend world-class courses like his, Thrun and two colleagues founded an online university called Udacity.[22] Udacity was soon followed by Coursera, EdX and many others. Access to knowledge, which for centuries was largely confined within the walls of academic institutions and available only to those who could afford tuition, was now available globally. Digital networks, led by the internet, are one major reason that the walls of the so-called "ivory tower" have fallen. This is not to suggest that there will be no future role for physically located universities and face-to face interaction among students and scholars. To the contrary, digital networks and tools extend, but do not replace, human senses and intelligence, especially considering the variety of linguistic, cultural and national contexts in which education takes place.

The importance of physical campuses and direct personal interaction in education is underscored by a related transformation – the rise of international branch campuses and education hubs in many parts of the world. Although some international branch campuses were opened as early as the mid-1950s, their growth in numbers began in the late 1990s and accelerated rapidly after the turn of the millennium, increasing from 84 at the end of 2000 to 249 at the end of 2015.[23] Some of these international branch campuses are located in so-called education "hubs" such as Education City in the UAE, Dubai International Academic City or the Incheon Global Campus in Korea.

ICT in Korean education

Given the important role of ICT in education, Korea gave top priority to the networking and computerization of South Korea's schools when it first began building its digital networks in the 1980s. Consequently, schools, colleges and universities throughout Korea, along with government organizations, were consistently among the first organizations in the country to enjoy the latest advances in high-speed networking and broadband internet as the country built out its infrastructure.

As the nation reached near-universal enrollment at the elementary, middle and high school levels, the placement of computer labs in those schools with high-speed internet also helped to minimize the problem of a digital divide. Children from families in all segments of society could enjoy the same level of computer and internet education at school.

When it came to the introduction of distance learning at the tertiary level, or the so-called cyber-universities, it came rather quickly and on a large scale. In 2001 and 2002 alone, there were 14 new cyber-universities founded. As of 2009 there were 18 cyber-colleges or -universities in South Korea with a freshman quota for admissions that year of 20,747 students.

As of 2009, online universities enrolled some 88,000 students, or 5 percent of the total number of university students nationwide. Many of the cyber-universities in Korea stressed lifelong education and offered courses in such niche markets as IT, real estate, counseling, nursing and education programs for the handicapped.[24]

Education in ICT: digital literacy and citizen engagement for the networked era

The public promotion of internet services and new digital media, also referred to as informatization, played a crucial role in ensuring that South Korea's new networks are used and are economically sustainable. The full benefits of ICT can only be realized when used for social means. This is the demand-side of the economic equation, and it is crucial to the building of an information society. It requires citizen awareness of both the nature of the information society generally and the uses of specific technologies that come with it. Unless citizens embrace the new ICT technologies, they will fail in the marketplace for shortage of consumer demand or politically for lack of support for the long-term commitment and investment that they demand.

At each stage of Korea's network modernization and infrastructure development, there was a broad and sustained emphasis on demand creation and public promotion of the information society. Informatization is to the information revolution what industrialization was to the industrial revolution. As noted by the World Bank Study, promotion of informatization requires both large-scale investment and the long-term cooperation of various organizations.[25]

Efforts to create such citizen awareness began in the early 1980s and have steadily increased over the years. Indeed, it is precisely the broad public acceptance of information culture, led by the internet that has fueled the rapid development of information technology in Korea. The idea of spreading or inculcating information culture in Korea included:

- changing mindsets and attitudes toward information use,
- developing a sound information environment and information ethics,
- enhancing citizens' capabilities to use information,

- spreading an information-centered lifestyle, and
- establishing relevant laws and regulations.[26]

In the 1980s, the Ministry of Communications emerged as a highly visible proponent of the information society and importance of information culture in everyday life. As the decade wore on, other ministries, including the Ministry of Science and Technology and the Ministry of Trade and Industry, also promoted this concept.[27]

Korea's informatization efforts proceeded in stages. They received a big boost with the establishment of the Ministry of Information and Communications and consolidation of its telecoms policy leadership in 1994 and the passage of the Informatization Promotion Basic Act in 1995.

From 2002–2004 special attention was paid to addressing the problem of digital divides, at both national and global levels. Both the "E-Korea Basic Plan" and the "U-Korea Basic Plan" were established, indicating a move from internet-based informatization toward the ubiquitous society. In January 2003, pursuant to Article 16 of the Act on the Digital Divide, the Korea Information Culture Center was upgraded to become the Korea Agency for Digital Opportunity and Promotion (KADO). In 2009 KADO merged with the National Information Society Agency (NIA), and in 2017 the NIA celebrated its 30th anniversary.

Perhaps most importantly, Korea's efforts to educate the public about ICT, also referred to as informatization, had solid financial support in the form of the Information Promotion Fund (IPF). The goals of the IPF included not only rolling out broadband networks, but also the promotion of e-government, support for ICT R&D, standardization and ICT education.

The IPF, based on both government and private sector contributions, allowed profits from the ICT fields to be reallocated into the ICT sector. From 1993 to 2002 the IPF reached U.S. US$7.78 billion. About 40 percent came from the government, 46 percent from private firms and 14 percent from miscellaneous profits and interest receipts.

A total of US$5.33 billion was invested between 1994 and 2003. Of that total, 38 percent was invested in ICT R&D, 20 percent into informatization promotion, 18 percent into ICT human resource development, 15.1 percent in broadband infrastructure and promotion, 7 percent in infrastructure in the ICT industries and 3 percent in standardization. In other words, if one includes R&D, fully three quarters of the IPF was used for educational purposes.[28]

Part of South Korea's immense educational effort over the years was directed toward technical training that would allow its citizens to use computers, the internet and the myriad of other technical gadgets that come along with modern digital networks. Without technicians to lay the fiber optic cable or to install the routers, switches and mobile base stations, one can hardly conceive of Korea today.

In 1985, the Information and Communication Training Center implemented programs for those graduating with non-computer majors and trained 1,900 people that year in ICT skills. By 1995 a total of 32,000 people had completed such training programs.[29]

The 1997–1998 economic crisis, most widely known in Korea as the "IMF Crisis," stimulated renewed attention to ICT training and did so on a massive scale. In 1999, on the heels of the crisis, the government announced a comprehensive informatization program aimed at improving the digital literacy of the entire Korean population. It was called, appropriately, "The Informatization Education Plan for 25 Million People," and its name indicated its ambitious character.

The plan aimed different education strategies at different target groups. Initially, it focused on 10 million students, 0.9 million government officials and 0.6 million of those serving in the military. This was later extended to include the disabled, housewives, the unemployed, farmers and fishermen. Broadcasting organizations were employed to educate people about IT communications with informatization education textbooks, and an information communication terminology book also being published and distributed.

Under this plan Information Education Centers were established at post offices throughout the country to offer education services to the general public. IT instructors were trained and supported through the Information Culture Center.

As if this were not enough, from 2000 through 2002 a separate basic information education plan was carried out side by side with the ongoing effort. This plan was aimed specifically at those who were socially disadvantaged and therefore had a lesser opportunity for education. Its aim was to minimize the digital divide within Korean society. By 2002 almost 11 million people, including housewives and farmers, had received ICT education. The program increased internet use among the population, helped develop the IT industry and aided expansion of the information infrastructure.[30]

The government's support of ICT education was not limited to technical aspects of the new digital media. Recognizing that digital contents add value to the nation's new digital networks, the government also undertook a range of efforts to support the development of new content. They included assistance to improve the education curriculum and increase the number of professors for digital content-related departments in universities and efforts to foster game developers. Government support for human resource development included measures to bolster education at home and abroad and to foster IT manpower that was actually demanded by industry.[31]

Globalization and study abroad in ICT development

The information revolution itself is a major factor contributing to the globalization of education, and Korean education has been profoundly affected by these developments. In the following pages we review some general developments in study abroad by Korean students, Korea's efforts to attract international students and the globalization of research and development.

The character of Korean migration changed decisively in the 1960s, when the country's economy began to develop and the government adopted an active emigration policy as a part of population control. Many Koreans moved to other countries in search of better economic opportunities, with the majority going to the United States. By 2003 there were 15 countries with more than 10,000

Koreans and five countries with more than 100,000. These latter five include the United States, China, Japan, the Commonwealth of Independent States and Canada, which together account for about 93 percent of overseas Koreans. Overseas Korean residents totaled 6.64 million in 175 different countries around the world, including 1.15 million temporary Korean expatriates. [32]

Academic studies have shown repeatedly that trade and investment bear an important relationship to diasporas. Our study adds support to that notion by showing the role of Korean engineers and academics in the U.S. in both the successful TDX electronic switching project and in jump-starting the semiconductor industry in South Korea. The business and social networks created by diasporas help to overcome informal trade barriers.[33]

During the early decades of South Korea's postwar development, while it was building its own system of university education, it began sending large numbers of students overseas to study. Their most popular destination, by far, was the United States. There are several aspects to this remarkable phenomenon.

First, the system of tertiary education in South Korea is broadly based on the American model. The academic calendar, courses offered, administration, activities and overall structure of admissions, grading and so forth all resemble colleges and universities in the United States.

Second, the popularity of study in the United States meant that Koreans who had earned Ph.D.s in the U.S. came to occupy positions of leadership in government, industry and academia. Today, three quarters or more of the professorships in leading South Korean universities are held by U.S.-trained Ph.D.s.

Third, study in the U.S. became so popular that South Korea became, for a number of years, the number one source of international students for U.S. colleges and universities. As of December 2008, over 110,000 Korean students were studying in the U.S., at all levels. This number exceeded the totals from Korea's more populous Asian neighbors China, India and Japan.[34] In recent years, this pattern has shifted. By 2018 as this is written, there were only 64,022 active students from Korea studying in the U.S. at all levels compared with 378,003 students from China and 227,199 from India.[35] Nevertheless, on a per capita basis South Korea still sends more students to the United States for study than any other Asian nation, including its largest regional neighbors.

Finally, it is important to underscore that study abroad is first and foremost a cultural experience. Those who went to the United States in the 1970s or even earlier, not only went from an East Asian to a Western culture, but from one of the poorest countries in the world to the world's richest country. This helps to explain the big impact the experience had on them. The cultural experience more than anything else was important.[36]

To be more specific, those who studied in technical fields like electrical engineering or other sciences that relate to ICT development, not only learned the subject matter being taught at their university, but had a chance to observe and participate in the American way of life. Such experiences ranged from using the telephone system in the U.S., with its collect calls, different style pay phones and other services that were not available in Korea at that time. For some of them, the contrast between relatively universal and reliable telephone service in the U.S.

and the utter lack of such service in Korea, imbued a strong determination to do something about it when they finished their study and returned to Korea.

Research by Choi showed that the Korean diaspora appeared to have a positive impact on trade by generating more exports than imports. He estimated that a doubling of the number of overseas Koreans appeared to increase South Korea's exports by 16 percent while increasing its imports by 14 percent. Significantly, his research underscored that the diaspora did not cause a brain drain but instead contributed to South Korea's development by transferring back home the knowledge and skills gained in more advanced countries.[37]

Innovation, research and development

Daniel Bell noted back in 1973 that "The joining of science, technology and economics in recent years is symbolized by the phrase "research and development" (R&D)."[38] Popular conceptions of R&D lead some people to think that it consists merely of scientists playing in their laboratories while using up large amounts of the government budget. Many think that scientists can squander hundreds of millions of dollars in citizens' taxes while only using their heads. To the contrary, research and development, properly understood, is a money-making business.

As shown in Chapter 2, two major R&D projects were at the heart of the epochal developments of the 1980s in South Korea. Without successful research and the development of the TDX switching system and the 4MB DRAM semiconductor, the telecommunications revolution of that decade might not have occurred. It was the success of those two R&D projects that ignited the information revolution in South Korea. One of their most important consequences was to instill an appreciation of the role of research and development among the nation's leaders.

Research and development activities in South Korea are conducted by private corporations, government research institutes and universities. As of 2006, 61 percent of all basic research and development in Korea was conducted by private corporations, with all but 7 percent of that being done by large firms.[39] Government research institutes accounted for 17 percent and universities 22 percent of basic research and development.

When compared with other countries in the world, Korea still has a relatively low percentage of R&D conducted in universities. As noted in 2009 OECD report, government research institutes were traditionally used as the vehicle to accelerate technology adoption, while universities were concerned with teaching.[40]

Business R&D (BERD) expenditures in Korea are heavily focused on high technology, especially ICT. One breakdown of BERD for 2004 showed that electronic parts accounted for 35 percent of total investment, while audio/video and communication equipment accounted for another 12 percent. Automobiles accounted for 15 percent and "other manufacturing" another 26 percent.[41]

Trends in Korean R&D

Korea has one of the world's highest levels of gross domestic expenditure on R&D (GERD). In 2006 it amounted to a little under US$30 billion, or 3.23 percent of

GDP. This was one of the highest levels in the world. Over the past several decades there have been several broad trends in R&D expenditure in South Korea.

First, as shown in Figure 5.1, South Korea has steadily increased its expenditure on research and development since the mid-1960s. This is represented by the line showing R&D as a percentage of GNP.

Second, there is a clear trend toward an increasing reliance on the private sector and a decreasing reliance on the government for R&D funding. The annual growth rate of business R&D in Korea is about twice the OECD average and is among the highest in the world. This reflects the emergence of Korean leaders in industrial technology, especially in information and communication technology, automobiles, shipbuilding and steel.[42]

Third, as Korea caught up with advanced countries and moved toward technological frontiers, it faced an increased need to conduct fundamental research. Accordingly, basic research increased from 13.6 percent of total spending in 1999 to 15.2 percent in 2006.[43] In particular, within private sector R&D the proportion of basic to applied research has increased. In 1998 basic research accounted for only 6.5 percent of private sector R&D, a figure that had increased to 12 percent in 2004.[44]

President Lee Myung Bak's government reorganization in 2008 shifted the government's R&D budget, which had been under the control of the Ministry of Science and Technology, to the new Ministry of Education, Science and Technology. The new government also declared that it would increase the level of government investment in R&D while placing more emphasis on basic or fundamental, rather than applied research.

As shown in Figure 5.2, South Korea's gross domestic spending on R&D as a percentage of GDP grew rather steadily from 1991 through 2016 except for a

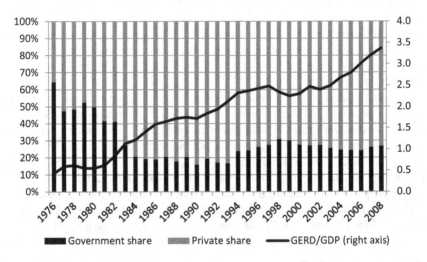

Figure 5.1 Gross expenditure on research and development in Korea, 1976–2008

Source: Ministry of Education, Science and Technology.

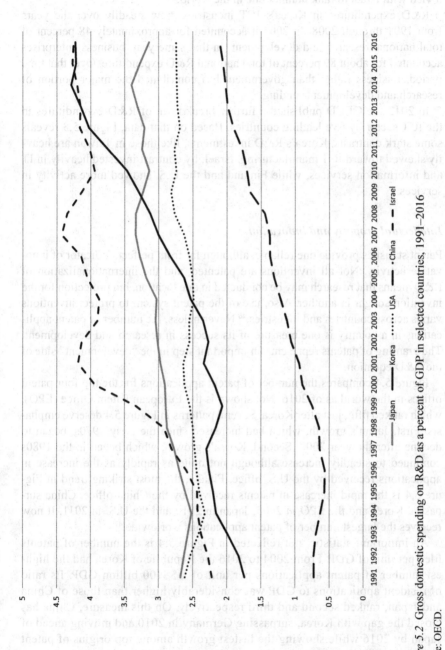

Figure 5.2 Gross domestic spending on R&D as a percent of GDP in selected countries, 1991–2016

Source: OECD.

Legend: Japan — Korea ⋯⋯ USA · China -- Israel

X-axis: 1991 1992 1993 1994 1995 1996 1997 1998 1999 2000 2001 2002 2003 2004 2005 2006 2007 2008 2009 2010 2011 2012 2013 2014 2015 2016

Y-axis: 0 0.5 1 1.5 2 2.5 3 3.5 4 4.5 5

decline in the wake of the Asian financial crisis in 1999. On this measure, Korea surpassed the United States in 2004, Japan in 2009 and Israel in 2014 after which it vied with Israel to rank number one in the world.

R&D expenditures in Korea's ICT industries grew steadily over the years from 1991 through 2008. In 2008 it accounted for approximately 48 percent of total national research and development. In that same year, business enterprises accounted for about 89 percent of total national R&D expenditure. Over that time period, business rather than government has contributed the major portion of research and development funding.[45]

In 2013 the OECD published a further breakdown of R&D expenditures in the ICT sector by five leading countries. Based on that data, Figure 5.3 reveals some stark contrasts. Korea's R&D investments, like those in Taiwan are heavily skewed toward ICT manufacturing. Israel, by contrast invested heavily in IT and information services, while Finland and the U.S. showed more activity in services.

Intellectual property and innovation

Patent statistics provide one reliable, although far from perfect, indicator of innovative activity. Not all inventions are patented, and the internationalization of R&D means that research may be conducted in one location, but protection for the invention sought in another. Also, use of the patent system to protect inventions varies across countries and industries.[46] Nevertheless, the number of patent applications in a country is one measure of its success in research and development. The granting of patents represents an important step in the "development" side of the R&D equation.

Figure 5.4 compares the number of patent applications for the top four patent offices in the world as of 2016. Not shown is the European Patent Office (EPO), which ranked fifth, just after Korea. Several patterns in Figure 5.4 deserve emphasis. First, Japan's growth, which had increased since the early 1970s, began to decline after the year 2000. Second, Korea's growth, which began in the 1980s continued to steadily increase although not quite as rapidly as the increase in applications received by the U.S. office. Finally, the most striking trend in Figure 5.4 is the rapid increase in patents received by the China office. China surpassed Korea and the EPO in 2005, Japan in 2010 and the U.S. in 2011. It now receives the largest number of patent applications worldwide.[47]

An important statistic, not reflected in Figure 5.4 is the number of patents filed per unit of GDP. From 2004 to 2016 the Republic of Korea had the highest number of patent applications per unit of US$100 billion GDP. Its ratio of resident applications to GDP was considerably higher than those of China and Japan, ranked second and third respectively. On this measure, China has closed the gap with Korea, surpassing Germany in 2010 and moving ahead of Japan by 2016 while showing the fastest growth among top origins of patent applications.[48]

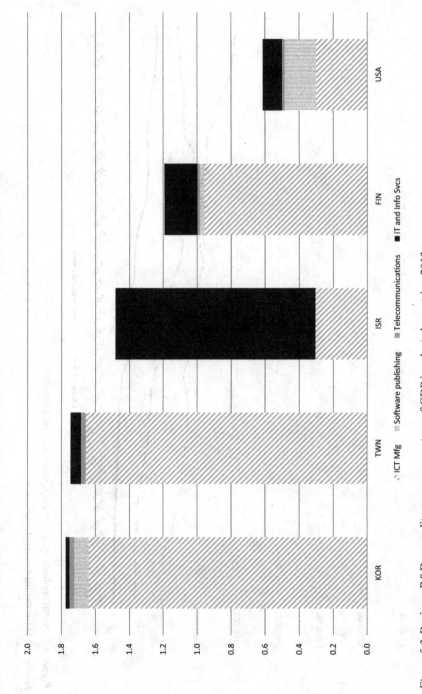

Figure 5.3 Business R&D expenditures as a percentage of GDP in selected countries, 2013

Source: OECD data.

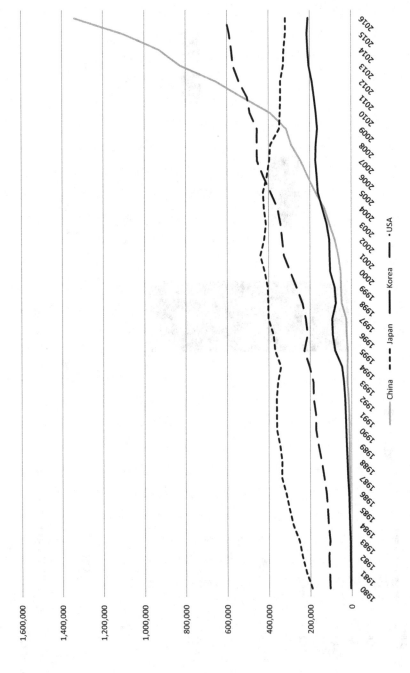

Figure 5.4 Patent applications by leading patent offices, 1980–2016

Source: World Intellectual Property Organization.

Globalization of ICT research and development

In recent years there has been a broad trend in ICT research and development toward more international and collaborative work. Many firms are embracing open innovation approaches and collaborating with external actors. Underlying these changes are the increasingly knowledge-driven nature of innovation and the changing organization of research and exchange of knowledge, driven by information technologies.[49] An OECD report noted that an increasingly globalized ICT R&D agenda is emerging with the following eight broad priorities:

- Physical foundations of computing
- Computing systems and architectures
- Converging technologies and scientific disciplines
- Network infrastructures
- Software engineering and data management
- Digital content technologies
- Human technology interfaces
- ICT and internet security and safety[50]

In 2005 the OECD sector including 21 ICT goods and services spent about two and a half times as much on R&D as the automotive sector and more than triple the pharmaceutical sector. By that time Korean ICT firms had caught up to firms in other advanced OECD countries in terms of R&D spending.[51] Notably, in 2007 Samsung Electronics overtook IBM in overall levels of R&D spending. By 2017 Samsung ranked third globally in R&D spending, following Amazon and Alphabet, the parent company of Google.[52]

Notes

1 Dahlman, Carl and Thomas Andersson, eds. (2000). *Korea and the knowledge-based economy: Making the transition.* The International Bank for Reconstruction and Development/The World Bank, and The Organization for Economic Cooperation and Development, pp. 14–16.
2 *Korea as a Knowledge Economy: Evolutionary Process and Lessons Learned.* Washington, DC: World Bank, conference edition, 2006, p. 2.
3 *Understanding Korean education. Vol. 5, Education and Korea's development.* Korean Educational Development Institute, 2007, p. 26.
4 *Understanding Korean education. Vol. 5, Education and Korea's development.* Korean Educational Development Institute, 2007, p. 5.
5 *Understanding Korean education. Vol. 5, Education and Korea's development.* Korean Educational Development Institute, 2007, p. 28.
6 Kim, Linsu. (1997). *Imitation to innovation: The dynamics of Korea's technological learning.* Cambridge, MA: Harvard Business School Press, p. 60.
7 OECD. (2008). *Education at a Glance*, p. 32.
8 Suh, Joonghae and Derek H. C. Chen. (2007). *Korea as a knowledge economy: Evolutionary process and lessons learned.* Korea Development Institute and The World Bank Institute, p. 41.
9 Note that this indicator does not provide information about the number of graduates actually employed and putting their skills to work.

10　Ministry of Education website. Retrieved November 2, 2018, from http://english.moe.
go.kr/sub/info.do?m=020105&s=english

11　*Education in Korea 2007–2008*. Republic of Korea: Ministry of Education and Human
Resources Development, p. 49.

12　WTEC report on the Korean electronics industry. *Executive Summary*, p. 8. Retrieved
from www.wtec.org/loyola/kei/welcome.htm

13　The Economist. (2005, September 8). The brains business. *The Economist*.

14　Benson-Armer, R. A. (2016, May). Learning at the speed of business. *McKinsey
Quarterly*.

15　Bailey, A. P. (2014). *The digital disruption of education publishing: How online learning is reshaping the industry's ecosystem*. Boston: The Boston Consulting Group.

16　Hennessy, J. (2002, September/October). The first great American university. *Stanford
Magazine*.

17　Hennessy, J. (2002, September/October). The first great American university. *Stanford
Magazine*.

18　Parker, G. G. (2016). *Platform revolution*. New York: W.W. Norton & Company.

19　Parker, G. G. (2016). *Platform revolution*. New York: W.W. Norton & Company, p. 6.

20　Parker, G. G. (2016). *Platform revolution*. New York: W.W. Norton & Company,
pp. 265–267.

21　MIT Opencourseware. Our history. Retrieved from https://ocw.mit.edu/about/our-
history/ Retrieved November 3, 2018.

22　Rifkin, Jeremy. (2014). *The zero marginal cost society*. New York: Palgrave MacMil-
lan, pp. 114–115.

23　Crist, John T. (2015). Innovation in a small state: Qatar and the IBC cluster model of
higher education. *The Muslim World*, 105, pp. 93–115.

24　Kang, Shin-who. (2009, July 22). Cyber universities solidifying network for growth.
The Korea Times. Retrieved from www.koreatimes.co.kr/www/news/special/2009/
10/242_48917.html

25　So, Chung-hae, Joonghae Suh and Derek Hung Chiat Chen. (2007). *Korea as a knowledge economy: Evolutionary process and lessons learned*. World Bank Publications,
p. 92.

26　*Korea's informatization policy to deliver ICT use in everyday life*. Seoul: Korea Agency
for Digital Opportunity and Promotion, 2007, p. 8.

27　See Chapter 5. Education, training and public promotion of information culture. In
James F. Larson, *The telecommunications revolution in Korea*. New York: Oxford University Press, pp. 149–169.

28　So, Chung-hae, Joonghae Suh and Derek Hung Chiat Chen. (2007). *Korea as a knowledge economy: Evolutionary process and lessons learned*. World Bank Publications,
p. 92.

29　*Korea's informatization policy to deliver ICT use in everyday life*. Seoul: Korea Agency
for Digital Opportunity and Promotion, 2007, p. 25.

30　*Korea's informatization policy to deliver ICT use in everyday life*. Seoul: Korea Agency
for Digital Opportunity and Promotion, 2007, pp. 21–22.

31　Broadband IT Korea: Connecting you to the digital world. *White Paper 2003*, Ministry
of Information and Communication, Republic of Korea, pp. 50–51.

32　Bergsten, C. Fred and Inbom Choi, eds. (2003). *Korean diaspora in the making: Its
current status and impact on the Korean economy*. Special Report 15, Washington,
DC: Peterson Institute for International Economics, pp. 16–17.

33　Bergsten, C. Fred and Inbom Choi, eds. (2003). *Korean diaspora in the making: Its
current status and impact on the Korean economy*. Special Report 15, Washington,
DC: Peterson Institute for International Economics, p. 19.

34　Student Exchange and Visitor Information System (SEVIS) General Summary Quarterly Review for the quarter ending December 31, 2008. Retrieved from www.ice.gov/
doclib/sevis/pdf/quarterly_report_january09.pdf

35 Retrieved November 17, 2018, from https://studyinthestates.dhs.gov/sevis-by-the-numbers.
36 Dr. Oh, Myung was one of these.
37 Choi, Inbom. (2003, January). Korean diaspora in the making: Its current status and impact on the Korean economy. Chapter 2 C. In Fred Bergsten and Inbom Choi, *The Korean diaspora in the world economy*. Special Report No. 15, Washington, DC: Peterson Institute for International Economics, p. 27.
38 Bell, Daniel. (1973). *The coming of post-industrial society*. Basic Books, p. 25.
39 *OECD reviews of innovation policy: Korea 2009*, Organization for Economic Cooperation and Development, 2009, p. 104.
40 *OECD reviews of innovation policy: Korea 2009*, Organization for Economic Cooperation and Development, 2009, pp. 78–79.
41 *OECD reviews of innovation policy: Korea 2009*, Organization for Economic Cooperation and Development, 2009, p. 105.
42 *OECD reviews of innovation policy: Korea 2009*, Organization for Economic Cooperation and Development, 2009.
43 *OECD reviews of innovation policy: Korea 2009*, Organization for Economic Cooperation and Development, 2009, pp. 74–76.
44 *OECD reviews of innovation policy: Korea 2009*, Organization for Economic Cooperation and Development, 2009, p. 103.
45 Oh, Myung and James F. Larson. (2011). *Digital development in Korea: Building an information society*. London: Routledge, p. 137.
46 *World patent report: A statistical review*, World Intellectual Property Organization, 2008 Edition, p. 10. Retrieved from www.wipo.int/export/sites/www/ipstats/en/statistics/patents/pdf/wipo_pub_931.pdf
47 *World intellectual property indicators*, 2017 edition. World Intellectual Property Organization, p. 32.
48 *World intellectual property indicators*, 2017 edition. World Intellectual Property Organization, p. 35.
49 Vickery, Graham and Sacha Wunsch-Vincent. (2009). R&D and innovation in the ICT sector: Toward globalization and collaboration. Chapter 1.8 In *The global information technology report 2008–2009: Mobility in a networked world*. World Economic Forum and INSEAD, pp. 95–110.
50 Vickery, Graham and Sacha Wunsch-Vincent. (2009). R&D and innovation in the ICT sector: Toward globalization and collaboration. Chapter 1.8 In *The global information technology report 2008–2009: Mobility in a networked world*. World Economic Forum and INSEAD, pp. 96.
51 Vickery, Graham and Sacha Wunsch-Vincent. (2009). R&D and innovation in the ICT sector: Toward globalization and collaboration. Chapter 1.8 In *The global information technology report 2008–2009: Mobility in a networked world*. World Economic Forum and INSEAD, pp. 97–99.
52 Retail News Asia. Samsung Electronics No. 3 globally for R&D spending. Retrieved from www.retailnews.asia/samsung-electronics-no-3-globally-for-rd-spending/

6 The global rise of Korea's electronics industry

International trade and exports

South Korea's digital development occurred during a period of transformation from the industrial economy of the 20th century to the globally networked digital economy. It coincided with the so-called "third wave" of globalization, which was driven by two main factors. The first was the technological change leading to lower costs for computing, communications and international travel that made it economically possible for firms to locate different phases of production in different and far away countries around the world. The second factor was the increasing liberalization of trade and capital markets.[1]

This chapter examines the remarkable growth of Korea's ICT sector over time and suggests where Korea fits in the context of global trends in the digital economy. It also documents the important role of Korea's large industry groups in its export-led development and the growth of these family-controlled conglomerates over time. It also examines the relative weakness of the service sector and how it led to the drive to develop a creative economy, with its emphasis on small and medium-sized enterprises and venture start-ups. Finally, the chapter will explore two promising growth industries for the future, robotics and online games.

Trade and innovation in Korea's digital development

Trade and innovation are increasingly recognized as necessary for sustained economic growth and prosperity for all nations, both developed and developing. As the foreword to a 2009 OECD-World Bank Study observed,

> We often think of innovation in terms of breakthrough inventions, but it can also be linked to organizational changes and technology diffusion. In a globalized world, in which countries and firms compete fiercely to buy and sell their products and services, innovation is a key driver of competitiveness.[2]

Noting South Korea's success in catching up with other OECD countries and reaching the technological frontier, the study benchmarked Korea to assess how well China, Brazil and other emerging economies were doing.

A key portion of the study, entitled *Innovation and Growth: Chasing a Moving Frontier*, argued that Korea's catch-up strategy involved a combination of interventions to promote export-led growth and support for innovative industries.

Growth in certain industries was powered by Korean exports. Between the early 1980s and 2004 the share of output exported increased from 38% to 64% in electrical machinery, which was Korea's leading industry, and from 5% to 33% in transport, its second largest industry. The overall share of exports in GDP increased from 23% to 43% between the 1970s and 2006.[3]

Korea's changing position in the global electronics industry

In order to place Korea's ICT sector and its growth in global perspective and to compare it with other countries, the sector must be clearly defined and measured. The OECD has played a lead role, along with the ITU and other international organizations, in defining and redefining the ICT sector. Since 2009 the ITU has published annual *Measuring the Information Society* reports.

The measurement problem is inherent in the nature of information and communications technologies. Since ICT production takes place in many industries, either as a principal or secondary output, it is not possible to get a complete measure of ICT production using industry statistics. The International Standard Industrial Classification (ISIC) system designed by the United Nations to classify economic activity was not created with the digital economy in mind. Nevertheless, identification of industries whose principal production is ICT goods or services is an essential component of a statistical framework to measure the ICT sector.[4] During discussions in 2006 the definition of the ICT sector was narrowed somewhat to the following: "The production (goods and services) of a candidate industry must primarily be intended to fulfill or enable the function of information processing and communication by electronic means, including transmission and display."[5]

As of 1985, the United States produced 45 percent of the world's electronics products, and Europe 21 percent. Japan's share was 24 percent, and it produced over 50 percent of the world's audio and video equipment.[6] Since that point in time East Asia has emerged as the world's dominant center of electronics production. By 2005 China and the nine major developing Asian countries accounted for 43 percent of worldwide production of electronics products.

As discussed below, the geographic shift toward Asia was accompanied by growth of trade, establishment of global production networks a shift away from OEM manufacturing and innovation, especially in key industry sectors. In addition, Korea's ICT industry itself experienced other dramatic changes, from the mid-1990s.[7] It saw the rapid rise of the telecommunications equipment sector and a shift in export destinations related to shifting production patterns.

Changing patterns of production and manufacturing

The growth of electronics production in Asia was accompanied by an equally impressive growth in electronics trade. The region's emergence went hand in hand with the development of global production networks in electronics. Most finished electronics products today are modular, and as costs of communication and transportation have decreased, flagship companies in the U.S., Japan and Europe, along with Korea's large companies, have fragmented their production chains vertically and offshored manufacturing activities to labor-abundant countries.[8] For example, in 2009 Samsung Electronics manufactured television sets or LCD displays at its manufacturing facilities in Mexico, India, Hungary, Indonesia, Slovakia and Brazil. In 2004 China overtook the United States to become the world's largest electronics exporter. However, this development was not fueled by the growth of Chinese companies, but rather by the relocation of production to China by multinational corporations in the West, as well as in Japan and Korea.

One result of this globalization is that it becomes more difficult to precisely credit one nation or one region with being the source of a particular product. Take for example the iPhone 4, launched in June of 2010. In Korea, media treatment of the launch featured extensive treatment of how the nation's leading handset manufacturers would be able to compete in the marketplace with various Android-based and other smartphones. However, the story soon came out that many of the most valuable components of the iPhone 4 are made in Korea. The actual assembly of all these modular components into a finished iPhone 4 was being done in China.[9]

Until the 1990s, the original equipment manufacturer (OEM) approach, in which a company makes parts or subsystems that are used in another company's end product, was the dominant pattern in South Korea's electronics industry. Korean manufacturers conducted joint ventures and concluded licensing contracts with such international companies as Philips, Micron, Intel, Toshiba, Sharp, Fujitsu, AT&T, NEC and others. OEM contracts were particularly important as a route of entry into the electronics industry because OEM clients provided guidance on technological and quality requirements for their products, as well as providing a market for the end products, allowing Korean companies to achieve economies of scale. The disadvantages of OEM arrangements included lower profit margins and a hindered ability to develop independent brand name recognition and marketing channels. The lower profit margins, in turn, made it difficult to make the R&D and marketing investments necessary to build their own brand products.[10]

Changing export patterns

By 2013 China had emerged as the world's top exporter of ICT goods, accounting for 32 percent of the total. It was followed by the United States with 9 percent, Singapore with 8 percent and Korea with 7 percent. Figure 6.1 shows the shifting pattern of Korea's exports from 1990 through 2015. It is based on WTO data for the following three commodity categories:

- Electronic data processing and office equipment.(SITC Division 75).
- Telecommunications Equipment (SITC Division 76). This category includes mobile phones, flat panel displays and television sets.
- Integrated circuits and electronic components (SITC group 776). This category includes the exports by Korea's semiconductor industry.

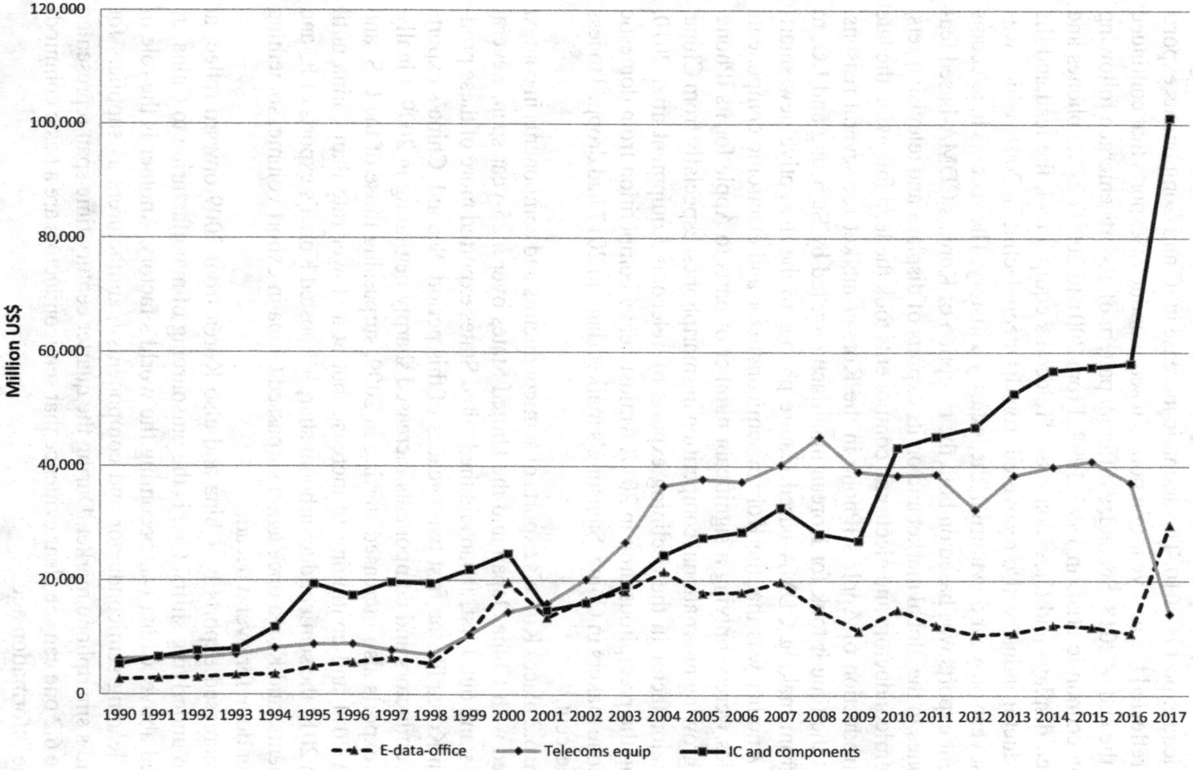

Figure 6.1 Korea's exports of electrical and electronic products.

Source: WTO Statistics on Merchandise Trade.

Million US$

120,000
100,000
80,000
60,000
40,000
20,000
0

1990 1991 1992 1993 1994 1995 1996 1997 1998 1999 2000 2001 2002 2003 2004 2005 2006 2007 2008 2009 2010 2011 2012 2013 2014 2015 2016 2017

E-data-office ▲ Telecoms equip IC and components

As of 2004, the three categories of exports together accounted for 12.2 percent of Korea's Gross Domestic Product.[11]

Figure 6.1 shows several distinct patterns. First, the export of semiconductors ("IC and components") led the way from 1994–2000 and again emerged as Korea's lead export in 2010. The dramatic increase in IC and Components exports in 2017 reflected the first year of a boom in semiconductor exports that continued through the end of 2018. Second, a different pattern of exports emerged following the Asian economic crisis in 1997–1998. From 2001 to 2009 mobile phones and flat screen displays and television sets ("Telecoms equip") took the lead and the semiconductor industry also recovered from its steep decline in 2001, which was the worst year-on-year downturn in the industry's history. The success in telecoms equipment exports was based on both the reception of Korea's CDMA-based feature phones in the global market and on its exports of displays and television sets. Third, semiconductors and electronic components took the lead following the late 2009 introduction of Apple's iPhone in the Korean market for several reasons. One was the time it took for Korean manufacturers, led by Samsung and LG, to switch from making CDMA-based feature phones to the disruptive new smartphones. Another was that the Korean semiconductor and electronic components companies were supplying a significant number of parts to Apple for its iPhone. Finally, growing international competition in smartphones, especially from China, was a major factor in the declining exports of Telecoms Equipment after 2010, including the precipitous drop in 2017. Amidst heavy competition from domestic smartphone makers in China, Samsung's market share in 2017 had dropped to less than 3 percent.[12]

Figure 6.2 places Korea's exports of semiconductors and components in context compared with China, Japan and the United States over a 25-year span. Several patterns stand out. First, Japan and the United States exported more of these products than Korea for the first two decades of this period. Second, China's exports of semiconductors and components increased sharply beginning in 2009. Finally, Korea's exports also increased starting in 2009, surpassing those of the U.S. and Japan in 2011 and continuing to increase, but at a slower rate than China until 2013. In 2017 the semiconductor boom sharply boosted Korea's exports of IC and components, and Korea once again surpassed China in export volume, something it had not done for over a decade.

The increased exports by China and also Korea from 2009 onward reflect at least two underlying shifts. One is the outsourcing of manufacturing to China, as discussed above. China was becoming the world's factory. Another is the role of Korea's leading semiconductor and components manufacturers as suppliers for the world's smartphone market. During the quarter century time span represented in Figure 6.2 one can see clear evidence that semiconductors are a basic commodity of the information age.

Figure 6.3 compares the same four countries on exports of telecommunications equipment. As with exports of semiconductors and components, it depicts the dramatic rise to global prominence of China. However, the increase in China's exports of telecommunications equipment started early and occurred even more

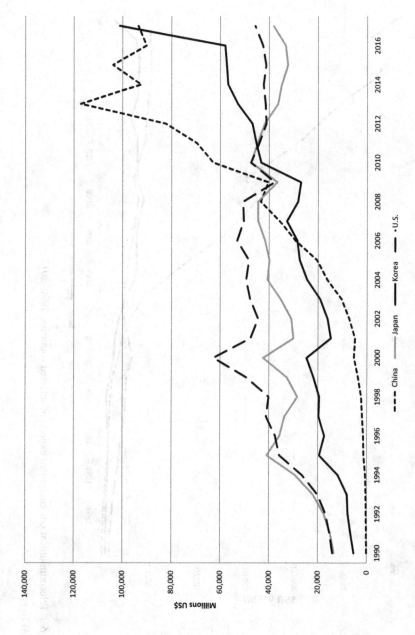

Figure 6.2 Integrated circuits and components exports by selected countries, 1990–2017

Source: WTO statistics.

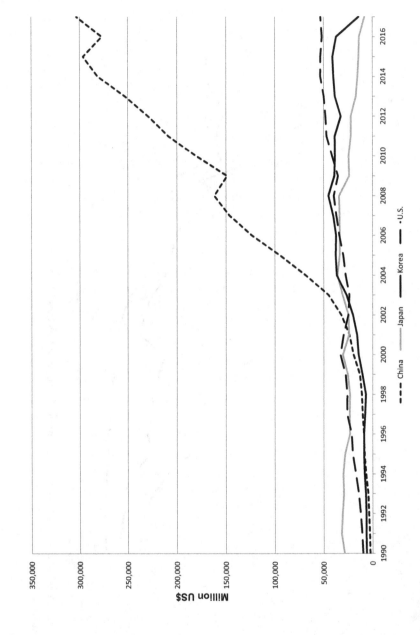

Figure 6.3 Telecommunications equipment exports of selected countries, 1990–2017

Source: WTO statistics.

rapidly. By 2002, Chinese exports in this category had surpassed those of Korea, the U.S. and Japan. Figure 6.3 also shows that Korea's exports of telecommunications equipment overtook those of Japan in 2004 and also led the U.S. from 2003–2009, after which the U.S. regained its lead. Finally, Figure 6.3 also reflects the sharp drop in Korea's smartphone exports in 2017, as discussed above.

The patterns represented in Figure 6.1, Figure 6.2 and Figure 6.3 illustrate both the risks and the rewards of Korea's heavy emphasis on manufactured ICT products and their export. The nation's leaders have long recognized that semiconductors were basic building blocks and a necessary commodity in the information age. In fact, as early as the 1970s, some referred to semiconductors as the "rice" for the future electronics industries. Given the importance of rice as a staple of the Korean agriculture this metaphor got the point across effectively. However, commodities are highly subject to the economic forces of supply and demand, with their price and profitability often running in cycles. Consequently, semiconductors were able to carry Korea's export-led economy during the 2017–2018 "super cycle" for memory chips when exports of smartphones and other electronics products declined.

Smartphones themselves are modular devices that rather quickly became commoditized. The ability of Chinese companies to obtain components at low prices and sell smartphones at lower prices than Samsung and other Korean companies goes a long way toward explaining the patterns discussed above. It also suggests that Korea's current technological lead in certain sectors, as discussed below, may not last long.

Developments in key manufacturing industry sectors

Korea is now a global leader in several key sectors of ICT manufacturing. These sectors include semiconductors, display and television sets, mobile handsets and the parts and components industry. A brief review of current issues in each sector underscores how Korea is now operating globally at the technological frontier in each.

In the semiconductor industry, South Korea's companies dominated the memory chip segments of the market. As of 2015 Samsung and SK Hynix were estimated to account for 70 percent of the worldwide DRAM market and 50 percent of the NAND flash memory market. In addition, Korea accounted for about 13 percent of the global LED market.[13]

The Korean semiconductor industry is shifting its focus from memory to non-memory or system-on-a-chip (SoC) semiconductors. This will help Korea's industry to provide total solutions for IT products. As the latest example of such an effort, Samsung provides Apple with the A4 chip that is used in its popular iPhones and iPad products. In June of 2010 Samsung Electronics announced that it would build a US$3.6 billion expansion of its Austin, Texas, chip plant to build a production line for system large scale integration (LSI) chips. The world LSI chip market was at that time dominated by Intel and was four times larger than the market for memory chips. Most of the chips produced on the new line would be for 3D television sets and mobile devices such as smartphones and tablet PCs.[14]

The display is a second sector of electronics manufacturing in which South Korea is a global leader, having overtaken Japan in 2002. As of 2014 Korea's share of the global market for display panels was about 43.5 percent.[15] The rapid development of this industry can partly be attributed to the fact that it achieved both the IT and mass production technologies during the period of transition from analog to digital technologies.

The display industry encompasses television sets, mobile phones, computer displays and screens of all shapes and size. The success of smartphones and Apple's iPad seems to suggest the logic of convergence toward a future in which "every surface will be a screen."

With rapid growth of the display industry, inbound foreign direct investment (FDI) by companies from advanced countries has increased, in recognition of South Korea's technology level. They include Merck, Toshiba, Asahi Glass and 3M, among others. This trend is expected to increase as the industry moves into next generation display models. Another example of the global scope of the display industry is an agreement between Apple and LG display. In January of 2009, Apple and LG Display announced a long-term agreement under which LG Display would continue to provide LCD panels for Apple's notebooks and monitors for the next five years. The agreement called for a US$500 million initial payment to LG Display.[16] LG also manufactured the display for the much-heralded iPhone 4, which prompted an interesting industry debate over which handset screen was superior, Samsung's AMOLED or the LG "Retina" screen used in the fourth generation iPhone.

In 2009 a Sungkyungkwan University-Samsung Advanced Institute of Technology (SAIT) team announced that it had discovered a manufacturing process for large-scale nanomaterial films that could herald the production of flexible electronic devices. SAIT said that the breakthrough would allow the country to make a grab for the global electrode market, which is critical in making displays, and could strengthen Korea's position in displays and semiconductors.[17]

The telecommunications equipment or apparatus industry constitutes a third manufacturing sector in which Korea is a global leader. As earlier chapters of this book made clear, the biggest single item in this industry sub-sector is mobile handsets. Korea's entry into the mobile handset market was made possible by its decision in the early 1990s to adopt CDMA technology. Its major industry players subsequently became a dominant force in the global marketplace for mobile handsets while manufacturing an expanding array of attractive "feature phones" that became popular in North America, Europe and all corners of the world.

The telecommunications equipment category also includes the base stations, switching equipment and transmission equipment that is necessary for both wired and wireless networks. Although less visible than the fashionable mobile phones, electronic switches and network equipment are form a very important part of Korea's exports.

Exports of communication equipment in 2007 stood at US$31.8 billion, up 13.4 percent from the previous year and with two thirds of that total being accounted for by mobile handsets.

Finally, the manufacturing part of South Korea's electronics sector today includes a healthy sector that manufactures parts and components. Harking back to the situation Korea faced in 1980 it is now possible to note the spectacular success the nation achieved in revitalizing its electronics sector. Today South Korea is a world leader in the production and export of electronic components for the IT industry. The old days when all the components had to be imported from more technologically advanced countries are over.

Software and content: apps, Google, games and TV

While manufacturing and export of ICT hardware has been one of Korea's traditional strengths, there is widespread recognition that future competitiveness rests increasingly with software and content. This becomes especially apparent with reference to the rapid developments in mobile broadband; the challenge posed by Google; and the future of such industries as online games, television and robotics.

As described in some detail earlier, the introduction of the Apple iPhone into South Korea's market came late compared with most other countries and created a distinct "iPhone shock." Part of this shock was the realization that the success of the iPhone had much less to do with the phone itself and everything to do with the number, variety and quality of the applications that consumers could utilize with the device. Indeed, the phone itself was manufactured in China, and virtually all of its most valuable components were made in Korea! Yet it was the software applications that caught the imagination of consumers and caused them to dramatically increase their use of mobile data services in Korea, as they had in other countries.

By all measures, Google shows up as the dominant search engine on the internet, worldwide. According to Netmarketshare, Google's market share of the global desktop search engine market is 84.8 percent. The same company estimates that Google's share of the mobile/tablet search engine market globally is 94.87 percent. In both estimates, Yahoo, Bing and Baidu make up the next three minor players in the overall global search engine market.[18]

In order to place Google's worldwide dominance in some perspective relative to Korea's internet activity, it is helpful to look at the overall internet audience worldwide. As of December 2008, China surpassed the United States and became the world's largest internet market with almost 180 million total unique visitors, as measured by Comscore. The United States had 163,300,000 unique visitors and was followed by Japan, Germany and the United Kingdom in rank order. South Korea ranked tenth with 27,254,000 unique visitors comprising 2.7 percent of the total worldwide internet audience.

However, American websites, led by Google, reach by far the largest internet audiences. Google sites reach 77 percent of the worldwide internet audience, with Microsoft ranking second at 64.2 percent, followed by Yahoo at 55.8 percent. The remaining top-ranked sites, except for China's Baidu, are mainly American social networking, shopping or media sites.

Google's significance extends far beyond its role as a search engine or web search portal. It is more like a global information utility, with the potential to be as dominant as AT&T, IBM or Microsoft once were.

Think for example of cloud computing. Although there are no official statistics on the number of servers in Google's cloud network, a Gartner report in 2016 estimated the number at approximately 2.5 million.[19] Google builds and continually updates what amounts to a virtual private network within the internet and completely interoperable with it.[20] As Google's bots crawl the internet, they are seeking to build a copy of as much of it as possible. The value added in its business model comes primarily from the value of search and of the applications that advertising revenues support.

Google's ventures into the provision of content also bolster its success in search and advertising. Since 2001 it has launched or acquired Google News, Blogger, Google Earth, Google Maps, YouTube, the Android platform, the Chrome browser, Google Voice and Google Books, to name some of its major ventures. In 2014, Google purchased a little known London start-up called DeepMind for more than US$600 million. At the time, few knew why Google made the purchase. One the reasons it turned out, was to further develop AI, including AlphaGo, an AI program that would eventually defeat the world's best Go player.

Google Books was launched in 2002. By 2003, Google had refined a nondestructive scanning process and resolved many tricky technical issues involved in scanning books in 430 different languages.[21] As of 2010 Google had scanned the contents of more than 12 million books.[22]

Google's accomplishments to date help to place in clearer focus the future of the internet and South Korea's potential role in it. One thing seems very clear. Language is both a limiting factor and an opportunity for Korea. While the home-grown search engine, Naver, does outstandingly well in the Korean-speaking market, it most likely will not export well. As Cyworld already experienced in trying to enter the U.S. market, it is extremely difficult to export web services that appeal to Korean linguistic and cultural tastes to non-Korean markets. However, message chatting apps such as Line of NHN and KakaoTalk are doing well. Kakao's new service, Kakao Taxi, is extremely popular in Korea.

However, Google's efforts to date also underscore the opportunity that presents itself. Its programs are available in many nations and many languages, seeming to underscore the vast multilingual nature of the internet and cyberspace. Google Korea could well partner with a Korean company to digitize all of the books in Korea and make them available to Koreans and Korean studies specialists at home and abroad. Google's encounter with the Korean government's requirement for real-name registration to upload video and comments on YouTube underscored the global nature of the internet but also illustrated business opportunities for Korea. In 2007 Korea's real-name internet registration system was expanded to include all websites with over 100,000 daily visitors. Google responded to this law by blocking YouTube users in Korea from uploading content. The issue was settled in 2012 when Korea's Constitutional Court overturned the real-name registration law, ruling that it violated the right to free speech.[23]

In one content sector, that of online games, South Korea has a running start and seems poised to be a world leader. That sector will be explored later in this chapter.

For most of the 20th century the United States, with Hollywood, has been the world's dominant source of films and television programs. For that reason, Noam has posed the question of whether internet TV will also be dominated by American content.[24]

In addition to online internet games, Korea has shown some strength in the animation market. However, even if we assume a growing global market for animations made in Korea, many of these will be custom-made for English or other language markets.

In an interview in late 2009, Eric Schmidt of Google suggested that within five years the internet would be a real-time, broadband intensive, video and app-centric web dominated by Chinese language content.[25] That prediction turned out to be not far from the reality. Although English remained the most widely used language on the internet as of June 2016, Chinese was in second place and closing in on English, based on the large worldwide population of Chinese language users, many not yet using the internet.[26]

Although future internet TV and video may be dominated by English and Chinese language products, there seems to be no reason, in principle, why Korean firms cannot thrive in that environment. The large population of overseas Koreans, in the United States, Europe and other parts of the world may indeed play an important role in building Korea's information society as the development of content and services become ever more critical.

Korea's global corporations

Korea today possesses a diversified industrial structure, in no small part because of its large family-controlled conglomerates. These are Korea's global corporations, led by Samsung, LG, Hyundai and SK, to name some of the leaders. Because the conglomerates tended to diversify into different industries, there are now large companies competing in several key industries, such as ICT, shipbuilding and automobiles. In fact, the large Korean conglomerates are generally defined as a parent company, owned or controlled by a family or extended family, that controls subsidiaries in various industries. As noted in an OECD study, the ICT industry is the most striking success story in Korea's industrialization. It became competitive and globalized in a relatively short period of time on the back of strong exports.[27]

Korea's large conglomerates have many affiliates. As of 2004, Samsung had 63, LG 46, Hyundai Motor 28, Hanjin 23, Lotte 36 and the list goes on. The structure of these conglomerates was cited by the IMF as one of the main reasons for the 1997–1998 Asian economic crisis. At that point they had diversified beyond their financial and technological capability, thanks in part to government protection. Consequently, they became a major target for reform. The government announced a number of requirements for corporate restructuring, including a focus on core businesses, the reduction of debt-to-equity ratios below 200 percent by 1999, dismantling of cross-credit guarantees among subsidiaries and management transparency.[28]

Companies like Hyundai, Daewoo, LG and Samsung helped to develop and commercialize key digital technologies over the years. As they grew, so did exports, the nation's GDP and the proportion of private sector contribution to Korea's national expenditure on research and development. The largest and most successful of the conglomerates also became large transnational corporations with facilities for research, manufacturing and sales straddling the globe. In terms of international recognition and brand value, Samsung achieved greater success than the others, especially in the ICT sector.

Leaders throughout government, industry and academia mostly agree that Korea must somehow reduce its dependence on hardware manufacturing and make a shift toward software and services. However, to date this has proven an elusive goal. The following concluding sections of this chapter look at two industry sectors in which SMEs already play an important role and that appear to be promising growth engines for Korea.

The robotics industry

By all measures, the robotics industry in Korea seems ready to mature into an important new growth engine for the Korean economy. This is not widely recognized around the world, leading the author of a 2014 article in the *Robotics Business Review* to label Korea as the "Quiet Giant of Asian Robotics."[29] The following are some milestones in the development of Korea's robotics industry and highlights of its growth.

Korea began designing and building robots in 1979, just on the verge of the "telecommunications revolution of the 1980s." The growth in use of industrial robots in Korea was largely driven by two major industry sectors: automotive and electronics. As shown in Figure 6.4, by 2014 this nation had by far the highest

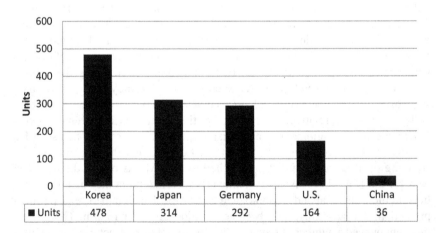

	Korea	Japan	Germany	U.S.	China
■ Units	478	314	292	164	36

Figure 6.4 Robot density in the five biggest destinations of robot supplies in 2014
Source: IFR World Robotics, 2015.

density of industrial robots in the world. As reported in the IFR World Robotics 2015 survey the average robot density worldwide for that year was 66. Robot density is the number of multipurpose industrial robots per 10,000 employees in the manufacturing industry.

In 2003 the Korean government identified robotics as a strategic national growth engine. A series of government initiatives followed, including:

- The 2008 Intelligent Robot Development and Promotion Act
- The 2010 establishment of a robotics division to oversee industry development in the Ministry of Trade, Industry and Energy
- The 2010 establishment of the Korea Institute for Robot Industry Advancement (KIRIA)

In addition, in 2008 the Korean government established a five-year US$1 billion program focused on supporting masters and doctoral students in robotics and mechatronics.[30]

For more than a decade, KAIST operated a robotics lab that developed Hubo, Korea's humanoid robot that won the 2015 DARPA challenge. The 2015 DARPA challenge pitted robots from top institutions and countries around the world in a challenge that focused that year on the successful completion of tasks in disaster response operations. The primary goal of the DARPA robotics challenge in 2015 was to develop human-supervised ground robots capable of executing complex tasks in dangerous, degraded, human-engineered environments. The 2015 DARPA challenge will be treated at greater length in Chapter 9.

In January of 2017 Korea's Ministry of Trade Industry and Energy (MOTIE) announced that it was upgrading KAIST's human robotics lab to an official government-supported research center that would receive approximately US$13 million dollars in funding over five years. The new focus of the lab would be on advanced mechanical engineering technologies and autonomous systems for humanoid robots.[31] The commitment of MOTIE to KAIST must be understood in the context of an earlier event that shocked Korea, the defeat of champion Go player Lee Sedol by Google's AlphaGo. As described in Chapter 4, the contest featured four separate matches. In the end, the computer defeated the human player four matches to one.[32]

The game industry

Globally, the game industry comprises an important and very interesting part of the apps ecosystem formed by IOS and Android. Apps are disrupting legacy industries in many sectors, including banking, retail and entertainment. In 2016 the global apps market grew 15 percent over 2015 and iOS App Store and Google Play revenues increased 40 percent. As of early 2017, the game industry accounted for 75 percent of overall revenue for apps on iOS and 90 percent of revenue on Google Play.[33]

As of 2016, South Korea was the world's second largest online games market after China. Its PC-based online games accounted for one fifth of revenue in the global market, but that was down from 28.6 percent in 2012. The Korea Creative Content Agency estimated that sales of Korean games abroad totaled about US$3.2 billion in 2015.[34]

Korea ranked fourth in the world as of 2015 in game revenues, after China, the U.S. and Japan, but to date has failed to release games with a strong impact on the mobile sector. This is significant since the whole global market is shifting away from the desktop games market toward games that can be played on mobile devices. As of 2014 South Korea's share of global game market for online games was 19.1 percent and its share of the global market for mobile games was 14.3 percent. Although game exports continued to increase between 2009 and 2015, their rate of growth slowed, due in part to the maturation of the large market in China. [35]

One of the reasons for South Korea's remarkable success in the game industry may well be that online games have their own language that is more or less universal. It doesn't require a great deal of text translation to play World of Warcraft, StarCraft or Lineage. Rather, there is more emphasis on visual symbolism and such universal themes as good versus evil.

Starting in 1998, the government encouraged national game companies to participate in well-known international gaming exhibitions. This eventually led 300 companies to advance overseas.[36]

In 2000, Korea hosted the World Cyber Games Challenge in Seoul. It attracted 168,000 participants from 17 countries playing 4 game titles. The WCG has now become widely recognized as a major international exhibition, attracting 1.5 million participants from 75 countries, competing in 12 game titles by 2007 and continuing to grow.

Since its inception in 2000 World Cyber Games has awarded more than US$4.6 million in prize money from 194 tournaments. Korea stood atop the country rankings based on winnings, followed by the United States, China, Germany and the Netherlands.

Notes

1   Soubbotina, Tatyana P. (2004). *Beyond economic growth: An introduction to sustainable development.* Washington, DC: The World Bank, p. 84.
2   Gurria, Angel and Robert B. Zoellick. (2009). Foreword. In Vanada Chandra, Deniz Erocal, Pier Carlo Padoan and Carlos A. Primo Braga, *Innovation and growth: Chasing a moving frontier.* OECD and The World Bank, p. 3.
3   Chandra, Vanada, Deniz Erocal, Pier Carlo Padoan and Carlos A. Primo Braga, eds. (2009). *Innovation and growth: Chasing a moving frontier.* OECD and The World Bank, p. 37.
4   OECD. (2011). *OECD guide to measuring the information society 2011.* Paris: OECD Publishing.
5   OECD. (2015). *OECD Digital Economy Outlook 2015.* Paris: OECD Publishing, p. 59.
6   Gangnes, Byron and Ari Van Assche. (2008, February). *China and the future of Asian electronics trade: Scientific Series.* Montreal: Cirano, p. 2.
7   Onodera, Osamu and Hann Earl Kim. (2008, September 26). *Case study 5: Trade and innovation in the Korean information and communication technology sector.* OECD Trade Policy Working Paper No. 77, pp. 16.

8 Gangnes, Byron and Ari Van Assche. (2008, February). *China and the Future of Asian Electronics Trade: Scientific Series*. Montreal: Cirano, p. 1.
9 iPhone 4 made in Korea. English *Chosun Ilbo*, June 10, 2010. Retrieved July 21, 2010, from http://english.chosun.com/site/data/html_dir/2010/06/10/2010061001059.html
10 Onodera, Osamu and Hann Earl Kim. (2008, September 26). *Case study 5: Trade and innovation in the Korean information and communication technology sector*. OECD Trade Policy Working Paper No. 77, pp. 12.
11 *The production of electronic components for the IT industries: Changing labour force requirements in a global economy*. Report for Discussion at the Tripartite Meeting on the Production of Electronic Components for the IT Industries, International Labour Office, Geneva, 2007, pp. 29–30.
12 Retrieved from www.counterpointresearch.com/china-smartphone-share/
13 SEMI. (2015, January 13). *The Korean semiconductor market*. Retrieved February 15, 2017, from Semi: www.semi.org/en/node/53891
14 Samsung plans US$3.6 billion expansion of Texas chip plant. *The Chosun Ilbo*, June 14, 2010. Retrieved from http://english.chosun.com/site/data/html_dir/2010/06/11/2010061100638.html
15 Invest Korea. (n.d.). *Displays*. Retrieved February 15, 2017, from Invest Korea: www.investkorea.org/en/world/display.do
16 Masterson, Michelle. Apple gives LG display a shot in the Arm. *ChannelWeb*, January 12, 2009. Retrieved from www.crn.com/hardware/212800117
17 South Korean scientists develop large film of nanomaterial to make flexible electronic devices. *Zoom Gadget*, January 17, 2009. Retrieved from www.zoomgadget.com/2009/01/south-korean-scientists-develop-large.html
18 Netmarketshare. (2017, January). *Mobile/tablet search engine market share*. Retrieved February 10, 2017, from Netmarketshare: www.netmarketshare.com/search-engine-market-share.aspx?qprid=4&qpcustomd=1
19 Retrieved November 20, 2018, from www.datacenterknowledge.com/archives/2017/03/16/google-data-center-faq
20 Cowhey, Peter F. and Jonathan D. Aronson. (2009). *Transforming global information and communication markets: The political economy of innovation*. Boston, MA: The MIT Press, p. 45. (Creative Commons version)
21 Google Books. Google books history. Retrieved from http://books.google.com/googlebooks/history.html
22 Open Book Alliance. How many more books has Google scanned today? Retrieved from www.openbookalliance.org/2010/02/how-many-more-books-has-google-scanned-today/
23 Choe, Sang-Hun. (2012, August 23). South Korean court rejects online name verification law. *The New York Times*. Retrieved November 20, 2018, from www.nytimes.com/2012/08/24/world/asia/south-korean-court-overturns-online-name-verification-law.html
24 Noam, Eli. (2004). Will internet TV be American? In Eli M. Noam, Jo Groebel and Darcy Gerbarg, *Internet television*. Lawrence Erlbaum Associates, p. 235.
25 Keen, Andrew. (2009, October 29). Google's ERIC Schmidt sets out the search engine's future. *The Telegraph*. Retrieved November 20, 2018, from www.telegraph.co.uk/technology/google/6459437/Googles-Eric-Schmidt-sets-out-the-search-engines-future.html
26 Internet World Stats. (2016, June). *Internet world users by language*. Retrieved February 10, 2017, from Internet World Stats: Usage and Population Statistics: www.internetworldstats.com/stats7.htm
27 *OECD reviews of innovation policy: Korea 2009*. Organization for Economic Cooperation and Development, 2009, p. 60.
28 *OECD reviews of innovation policy: Korea 2009*. Organization for Economic Cooperation and Development, 2009, p. 61.

29 Edwards, J. (2014, May 8). The quiet giant of Asian robotics: Korea. *Robotics Business Review*.
30 Edwards, J. (2014, May 8). The quiet giant of Asian robotics: Korea. *Robotics Business Review*.
31 Asian Scientist (2017, January 25). US$13 million humanoid robot research center opens in South Korea. Retrieved from: https://www.asianscientist.com/2017/01/tech/humanoid-robot-research-center/
32 Cho, M. H. (2016, March 15). *Google AlphaGo caps victory by winning final historic Go match*. Retrieved February 10, 2017, from ZDNet: www.zdnet.com/article/google-alphago-caps-victory-by-winning-final-historic-go-match/
33 Takahashi, D. (2017, January 17). *App Annie: Worldwide app downloads grew 15% and revenue soared 40% in 2016*. Retrieved February 10, 2017, from VB (Venture Beat): http://venturebeat.com/2017/01/17/app-annie-worldwide-app-downloads-grew-15-and-revenue-soared-40-in-2016/
34 Song, J. A. (2016, August 28). South Korea gaming groups struggle to fend off China. *Financial Times*.
35 Song, J. A. (2016, August 28). South Korea gaming groups struggle to fend off China. *Financial Times*.
36 *2007 the rise of Korean games: Guide to Korean game industry and culture*. Ministry of Culture and Tourism, Korea Game Industry Agency, p. 7.ss

7 Energy and ICT
Transition to green growth and sustainable development

As discussed in the opening chapter, the context for the digital network revolution that originated in the mid-20th century included globalization, urbanization and growing scientific evidence and public concern about environmental sustainability. This chapter addresses the relationship of the rapidly emerging and changing digital ecosystem to the environmental ecosystem of planet earth. The relationship of the ICT sector to the challenge of sustainable development, globally and in Korea, deserves careful scrutiny for several reasons.

First, the ICT sector itself is at once both a contributor to environmental degradation and perhaps the best hope for dealing with it on a global scale. The massive data centers that allow cloud-based computing via the internet consume large amounts of energy, contributing to the carbon footprint of corporations and nations. One empirical study of the carbon impact of the whole ICT industry included PCs, notebooks, monitors, smartphones and servers. It found that the ICT industry's contributions to worldwide greenhouse gas emissions (GHGE) would rise from 1–1.6 percent in 2007 to 3–3.6 percent in 2020. Assuming continued relative growth, ICT's contribution to global GHGE would exceed 14 percent by 2040. The study suggested that communication infrastructure, including networks (24 percent) and data centers (45 percent) together would account for nearly 70 percent of the ICT footprint. In addition, it estimated that smartphones would contribute about 11 percent to the total ICT footprint by 2020, exceeding the contributions of desktops (6 percent), laptops (7 percent) and displays (7 percent). The study concluded that the short average useful life of smartphones, along with the energy required for their manufacture, were major elements of an unsustainable business model.[1]

On the other hand, the ICT sector holds the promise that through appropriate application of tomorrow's increased storage, computing and communications capabilities, both human and artificial intelligence might be applied to smart grids, public safety networks and many other efforts to achieve sustainable development.

Second, Korea presents a cogent example of the relationship between the rise of the network society and the shift, in public policy terms, from a brown growth to a green growth strategy. As argued in this book, the ICT sector became the engine of South Korea's growth in the 1980s, and the nation is regarded as an outstanding example of ICT-driven development. However, it was not until 2008 that the nation turned in a new direction, adopting an aggressive green growth strategy as a matter of national policy. Why and how did this shift in policy take place and what was the role of the ICT sector in it?

Finally, an exploration of such questions promises to further clarify lessons that might be helpful for policymakers in other countries. As discussed in detail in Chapter 9, Korea is a leader in next generation networks and a testbed for the world. It also plays these roles in relation to the smart grid industry and other cutting-edge efforts to develop clean, sustainable energy. Given the nation's world leadership in broadband networks, this chapter places Korea in the context of national energy transitions in other countries around the world and its role as an aid donor and exporter of sustainable energy technologies.

Environmental sustainability in the digital age

The 1987 Brundtland Commission Report, formally named the *Report of the World Commission on Environment and Development: Our Common Future*, is widely considered to mark the point at which environmental sustainability became part of the global policy agenda. The Commission was formed in late 1983 at the request of the UN Secretary General and was chaired by Norway's Prime Minister Gro Brundtland. It offered a succinct definition of sustainable development: "Humanity has the ability to make development sustainable to ensure that it meets the needs of the present without compromising the ability of future generations to meet their own needs."[2]

While many elements of the Brundtland Commission report remain relevant today, two critical sets of issues are in much sharper focus as this is written, three decades later. One is the challenge of sustainability in economic, social, political and environmental terms. The second set of issues centers on the economic, social, political and cultural transformations and possibilities presented by the internet and ICTs. Souter suggests that a central question arises from the juxtaposition of these sets of issues: "How far and in what ways do we need to change our understanding of sustainability in the light of the information and communications revolution?"[3]

A review of the evolution of thought about sustainability is beyond the scope of this chapter. However, as noted in Chapter 1, the concept of sustainability has traditionally been thought of in terms of its social, economic and environmental dimensions, represented as intersecting circles. More recent analyses suggested that two additional dimensions might be added to understand sustainability. One is cultural diversity, understood in relation to globalization of communications, economy and society and the more intensive intercultural interactions that take place as a result. The other is governance, referring to the institutional mechanisms, rules and norms that shape decision-making and behavior by governments, businesses and citizens.[4]

We would argue that both cultural diversity and governance issues, along with the social, economic and environmental aspects of sustainability are profoundly shaped by the digital network revolution. The new digital networks dramatically extend the scope for social, political and economic interaction and for humans to address the challenges of sustainability. Thanks to advances in ICTs, the new networks extend globally and cut across all industries and economic sectors with a potentially profound impact not only on economics, but also on politics, society and culture. As Souter put it, "The rapid development of information and communication technologies and their increasingly pervasive influence on human activity have added to this complex and evolving context for sustainability."[5] Among the changes in information and communication technologies that took place since the first Earth Summit, two are particularly noteworthy: the adoption of the internet and mobile phones. However, he notes earlier in the same report that individual elements in the sustainability framework shown in Figure 7.1 have frequently been pursued by governments and other organizations in silos or in a manner that was detrimental to the pursuit of other elements.[6]

Sustainability in the network society

In historical perspective, the digitization of telecommunications networks and the emergence of what is variously termed an information society, knowledge economy or network society coincided with the first effort to define sustainable development and subsequent global efforts to deal with the problem. In the short span of about three decades since the Brundtland Commission report, business,

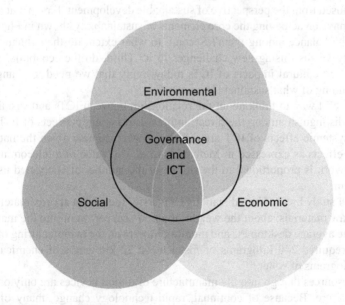

Figure 7.1 Dimensions of sustainability in the digital era
Source: Authors' adaptation from various sources.

politics, economics and society in general experienced the disruptive influence of new digital technologies, including the following:

- The internet
- Mobile telephony and mobile internet access
- Social media and networking
- Cloud computing
- The Internet of Things
- Big data analysis and visualization

As general purpose technologies, ICTs have sweeping effects on economies and societies, disrupting many patterns of activity that were common in the 20th century industrial era in society, politics and economics. Indeed, the disruptive or revolutionary character of these impacts is one reason why some people are still skeptical that the information society, which is more aptly called the network society, has arrived.[7] In point of fact, we agree with Souter that the changes brought about by the digital network revolution are so central to societies, economics, politics and culture in the 21st century that they force a rethinking of the concept of sustainability itself.

The potential influence of ICTs on sustainable development may be thought of as a double-edged sword. While the new digital technologies have unquestionably brought benefits to many people, they have also no doubt contributed to unsustainable patterns of behavior such as demand for non-renewable energy. Given this double-edged nature of the digital technologies per se, three general questions may be posed from the perspective of sustainable development. First, what impact do ICTs have on achieving the core elements of sustainability shown in Figure 7.1 and on the balance among them? Second, to what extent do they enhance sustainability versus raising new challenges to it? Third, do the economic, social, political and cultural impacts of ICTs today imply that we need to change our understanding of what sustainability means?[8]

One useful way to think about the relationship between ICTs and sustainability is to distinguish among the direct, indirect and systemic effects of ICT.[9] The overall systemic effects of ICT are difficult to overestimate given the nature of network effects as expressed in Metcalfe's Law. The value of a telecommunications network is proportional to the square of the number of connected users of the system.

A 2004 study by the United Nations University found that approximately 1.83 tons of raw materials, about the weight of a mid-size car, went into the manufacture of the average desktop PC and monitor. Moreover, the manufacturing of these devices required 240 kilograms of fossil fuels, 22 kilograms of chemicals and 1,500 kilograms of water.[10]

The resources that go into the manufacture of digital devices are only one part of the picture. Because of continual, rapid technology change, many of these devices have a relatively short life. Smartphones provide an excellent example of this phenomenon and one with which most Korean consumers are very familiar.

The increasing generation of e-waste is a growing problem, calling attention to problems ranging from the leaching of chemicals and metals from landfills, to working conditions in e-waste recycling plants located in developing countries.[11]

Overall, the impact on sustainability from the use phase of ICT accounts for two thirds or more of the energy used, greenhouse gases created and waste generated. In the EU as of 2010, the average lifetime of a PC was 3 years and that of a mobile phone 18 months. Estimates of the total amount of e-waste or computers thrown away each year in the EU and globally are staggering.[12]

From millennium development goals to sustainable development goals

A report compiled by researchers from The Earth Institute at Columbia University in collaboration with Ericsson notes that the Sustainable Development Goals (SDGs) seem utopian to many and acknowledges that they are "stretch" goals that require a transformation of societies that is far deeper and faster than in the past. Business as usual (BAU) will not be sufficient for success. However in their view, "the broad application of information and communication technology (ICT) is a profound reason for optimism, since the rapid development of ICT-based services and systems offer the possibility for the needed deep transformation of the world economy and societies more broadly."[13]

The optimistic view of Jeffrey Sachs and colleagues at The Earth Institute is grounded in a belief that "In essence, ICTs are "leapfrog" and transformational technologies, enabling all countries to close many technology gaps at record speed." They cite the unexpectedly rapid and near-universal uptake of mobile telephony in support of this view, but note that "This transformation needs to be scaled up. It is also our assertion that future advances in ICTs – including mobile broadband, the Internet of Things (IoT), robotics and artificial intelligence, 3-D printing, and others – will provide the tools for additional, unprecedented advances in healthcare, education, energy services, agriculture, and environmental monitoring and protection."[14]

The ICT portfolio

The SDG ICT Playbook, a document produced as a joint effort of industry with leading NGOs,[15] suggested that the main technology building blocks for use in achieving the sustainable development goals can be represented as a portfolio of the following technologies and services, all drawing on the new power of cloud computing:

- Analytics – turn vast amounts of geocoded, structured and unstructured data into actionable information and knowledge.
- Social media – connect people to people across the globe.
- Digital services – connect people to information and allow people to engage, share and transact, regardless of location.

- Smart systems – generate efficiencies by automating work, responding to events that impact that work and optimizing the use of resources.
- Satellites and Unmanned aerial vehicles – put information in the context of the planet on which we live.
- The Internet of Things – connects anything from sensors to intelligent devices to people and systems over the internet to support evidence-based decision-making.
- Connectivity – provided by data and telecommunication networks connects people to people, information and digital services.
- Mobile devices – allow people to stay connected and take advantage of ICT solutions anywhere and anytime.
- Power – and the innovative approaches for providing it ease the deployment of technology even in remote locations.
- 3D printing – enables the production of objects such as tools and spare parts on demand from any location

Global citizenship education and engagement

At its heart, education involves choices about what to study, research or teach. The curriculum and content of education is integral to its prospects for success. As noted earlier, the development of computers and digital networks led Korea to conduct massive campaigns to educate citizens about the new technologies. In addition to technology education and the whole range of traditional academic subjects, a new topic has come more clearly and urgently into focus since the first edition of this book was published in 2011. It is the challenge of environmental sustainability.

Work by researchers at the International Institute for Sustainable Development (IISD) in Canada frames the issue clearly. They identify two issues of profound importance to the future of global economies and societies. One is the challenge of environmental sustainability, which amounts to the future of planet earth as we know it. The other is the potential of ICT.[16]

The IISD framing of the issue is succinct and persuasive. The rapid development of digital technologies (ICTs) makes it difficult to imagine a sustainable future without leveraging the possibilities presented by the new technologies. Likewise the new technologies can only thrive if the planet is sustainable.

Inevitably sustainable development demands global citizenship education and engagement. In a luncheon speech at The 3rd UN Academic Impact Seoul Forum on Citizenship Education, Professor Young Soogil clearly articulated what is at stake. He argued that educators at institutions of higher education have an obligation to the younger generations.

> I argue that you, the educators at the higher education institutions, should engage your students in learning about the problem of climate change, how countries should work with one another in order to cooperate to solve the

problem, and what could be the solutions, and ask your students how, as individuals, they should contribute to the solution of this collective action problem.[17]

In this way, he emphasized, students might learn what it means to become a good global citizen.

In 2008, as a matter of national policy, Korea announced a dramatic shift from the brown growth strategy it had pursued to one of green growth. Chapter 7 will examine in more detail that transition and the role of digital technologies in green growth.

Korea's policy shift from brown to green growth

After half a century of fossil-fueled "brown growth," Korea's shift to a green growth as a long-term national policy objective was remarkably abrupt. From 1990 to 2007 the nation's carbon emissions grew faster than any other nation in the OECD, making it at that time the ninth largest carbon dioxide emitter in the world. Given Korea's heavy dependence on manufacturing industries and exports, it was not difficult to understand its historic reluctance to sign such international agreements to reduce carbon emissions as the Kyoto Protocol and Copenhagen Agreements.[18]

Korea's green growth initiative stands out among similar efforts by other countries for its ambition, speed of deployment and systematic design and execution. Kim and Thurbon explain it as "developmental environmentalism" and suggest that it embodied the spirit of "race to the swift" that defined Korea's approach to techno-industrial transformation since the 1960s. They argue that Korea's prioritization of green growth in 2008 reflects the reimagining of the relationship between the economy and the environment on the part of the policymaking elite. The result of this reimagining is the philosophy they call "developmental environmentalism," which draws heavily on the main tenets of developmental state theory.[19]

The evidence suggests that the primary motivation for Korea's sudden embrace of green growth was the nation's extreme dependence on fossil fuel imports. When President Lee Myung Bak took office Korea was importing 97 percent of its total energy requirements, making it highly vulnerable to international energy price fluctuations. The global price of oil went from US$30 to over US$100 between 2003 and 2007, and in 2008 it rose to almost US$150 per barrel.[20]

The Lee Myung Bak initiatives

Lee Myung Bak won the December 2007 presidential election and formally took office in February of 2008. His business career included long stints as CEO and Chairman of Hyundai Construction. He served as mayor of Seoul from 2002–2006, during which his signature accomplishment was restoration of the Cheonggyecheon stream.

Upon his election Lee Myung Bak, like all of his predecessors as president of the Republic of Korea, began searching for a project that could brand his administration's policy and also provide a new growth engine for the Korean economy during difficult times. The environmental concern was not at the top of President Lee's mind, but he was heavily influenced while mayor of Seoul by Dominic Barton, who headed the McKinsey Korea office at that time and later became Global Managing Partner at McKinsey. Barton advised Lee Myung Bak that the environmental agenda was the global agenda of the times.

After he was elected, President Lee assembled a team of strategic thinkers to consider what his administration's priority projects should be. Among Lee's senior staff was a former journalist from the *Maekyung Business Daily* named Kim Sang Hyup. Kim had been in charge of long-term vision projects and related conferences while with *Maekyung*. Dominic Barton put the green growth idea into President Lee's mind, and Lee instructed Kim Sang Hyup to pursue the idea.[21]

The idea developed steadily during the first few months of the Lee Myung Bak administration. President Lee first used the term "green growth" in an address on the national Independence Day holiday in August of 2008 when he proposed green growth as Korea's new long-term developmental strategy. Immediately after the August speech, President Lee appointed Kim Sang Hyup as secretary of the environment and instructed him to lead formation of a Presidential Committee on Green Growth. Kim was also instructed to prepare a long-term strategy document covering the period up to 2050. There were about 20 meetings of the Presidential Committee on Green Growth during Lee's tenure as president and he attended most of them, missing meetings only when traveling abroad. The committee had 36 official members plus about 15 other participants. The committee meetings were co-chaired by the Prime Minister and Dr. Young Soogil. Typically heads of relevant organizations, private business representatives and senior staff of related ministries were invited to attend comprising a total attendance of two hundred or more people at the peak.[22]

Following President Lee Myung Bak's August 2008 green growth address, four unique milestones propelled the nation's pursuit of green growth[23]:

- building the solid governance framework for green growth by establishing the Presidential Committee on Green Growth (PCGG) in 2009;
- strengthening the legal enabling environment for green growth by enacting the Framework Act on Low Carbon Green Growth in 2010;
- mobilizing various ministries to formulate comprehensive green growth plans at various levels – sectoral, national and local – including the National Strategy for Green Growth (2009–2050) and the Five-Year Plan (2009–2013); and
- honoring Korea's commitment in the global climate change agenda by setting an ambitious GHG reduction target of 30 percent by 2020, the highest recommended target for a non-Annex 1 country of the Kyoto Protocol.

Despite such legal and institutional measures President Lee Myung Bak liked the green growth initiative primarily because it provided a growth and development strategy, not because it reflected a longer-term or deeper commitment to sustainable development. Furthermore, Lee's successor as President, Park Geun Hye, downgraded the Presidential Committee on Green Growth to a Prime Ministerial committee and did not even convene it until eight months into her term. During her nearly five-year term in office, President Park rarely uttered the term "green growth," having chosen the "creative economy" as her signature project.[24]

Along with green growth, the Four Rivers Project became a major legacy of the Lee Myung Bak administration. President Lee knew that water supplies would be a future problem for all countries, so he implemented a project to increase the water holding capacity of South Korea's major rivers by dredging them and creating holding areas. Unfortunately, this damaged Korea's river ecosystem and contributed to the problem of green algae blooms. The destruction took place to an extent that President Lee Myung Bak's signature green growth initiative was called "anti-environmental."[25]

Market liberalization

Introduction of competition into the power market began in 2001 when the government allowed companies other than KEPCO to produce power under a liberalization program. Since then these independent power producers, which include Posco, SK and GS, can sell their power through the Korea Power Exchange (KPX). They accounted for 25 percent of total electricity capacity in 2016, and the government plans to increase their share to 34 percent by 2029.[26]

KPX was established under the Electricity Business Act of Korea as of April 2, 2001. It was designated as the core organization responsible for management of the newly introduced competitive electricity marker. At the same time, six electricity-generating subsidiaries were separated from KEPCO in a reorganization of the electricity supply sector in Korea. As of that date the supply sector included KEPCO (transmission and distribution, monopoly wholesale purchaser and retail supplier) and the six newly established generating companies (GenCos) and existing independent power producers (IPPs).[27]

A study published in 2016[28] argued that liberalization of the electricity industry is key to implementing the Korean government's plans for nationwide deployment of smart grids by 2030. The following section explores in more detail Korea's experience with and plans for smart grids.

The smart grid in Korea

A "smart grid," almost by definition involves the introduction of digital network capabilities into the power grid, which until recently has been a one-way network for the production and delivery of electric power. Consequently, it should hardly come as a surprise that Korea seeks to become a leader in smart grids. Because the nation possesses more robust and developed broadband networks than other countries, Korean smart grid and IoT initiatives are able to take backhaul connectivity for granted.

Kim[29] argues that Korea's approach to building smart grid infrastructure involves the creation of a new "hybridized industrial ecosystem" (HIE) in which the hybrid character refers to new forms of government-industry collaboration. The HIE leverages the capabilities of Korea's large industrial conglomerates and their networks of small and medium-sized industry suppliers. Kim suggests that Korean policymakers, like their counterparts in Taiwan, view smart micro-grids as a new developmental infrastructure and export growth engine. The HIE strategy links all elements in the innovation value chain and aims at exporting complete technology systems and solutions, not just individual components as in earlier eras.

The Korea Smart Grid Association (KSGA) and its standards setting arm, the Korea Smart Grid Standardization Forum (KSGSF) are two key institutions driving the nation's smart grid initiatives. Representing all key players in the power systems, energy and green ICT industries 143 companies and research institutes are members of the KSGA.[30]

The smart grid initiative in Korea outlined a platform for completely rethinking the electricity grid to make it viable in the 21st century. The five key components of this platform are as follows:

1   Smart power – Intelligent monitoring of the demand, high level of fault tolerance and fast restoration in case of failures. In the UK we would add the ability to import large quantities of power from interconnectors, turned on or off with very short notice.
2   Smart service – Providing domestic, commercial and industrial customers with electricity tariffs and services tailored to their needs, incorporating the ability to verify test data and to flow power in two directions everywhere.
3   Smart place – Allowing the introduction of intelligence in the home, particularly in major appliances that generate most of the demand; real time pricing and demand management.
4   Smart transport – Sophisticated systems for managing the connection of massive numbers of electric vehicles to the grid so that their demand is met without overwhelming the system.
5   Smart renewables – Connecting a large number, and diverse set, of variable sources of generation to the grid while maintaining high levels of stability.[31]

There are three types of smart microgrids under development in Korea: an urban-based "Smart Grid Station (SGS)," "Energy Block Platforms" suitable for industrial complexes and remote island-based microgrids.[32]

Urban smart grid stations

Based on experience gained from the Jeju Smart Grid project, which began in 2009, KEPCO built a smart grid station (SGS) in its Guri branch building in Seoul. The first SGS was designed as a demonstration project in which "station" refers to a place or building capable of providing intelligent electricity services to customers. These services would be different from existing building energy management (BEM) systems. The Guri demonstration project met or exceeded its targets for energy-saving, which led KEPCO to install SGSs in 121 of its branch offices. The demonstration also showed that this building-centered approach could be applied to other buildings.[33]

Energy block platforms

Industrial complexes and special economic zones played an important role in Korea's economic development over the past several decades. However, the nation's rapid development caused severe environmental degradation and the industrial complexes that were once the symbol of the Korean economic miracle became areas to avoid because of various forms of pollution. In 2003 the Korean National Cleaner Production Center (KNCPC) launched the national Eco-Industrial Park (EIP) program as part of efforts by MOTIE to promote industrial development that could simultaneously achieve environmental sustainability.[34]

The number of industrial complexes in Korea grew from two in the 1960s to 1,124 in 2015. Approximately 80,000 companies operated in various industrial complexes, and their economic output totaled US$929 billion in 2015, representing 63 percent of Korea's GDP. The industrial complexes created wealth and contributed greatly to Korea's rapid urbanization. However, they also caused considerable harm to natural ecosystems and also to the general well-being of local communities. In Ulsan, pollutants from industrial zones contaminated air and water, making it one of the most unlivable cities in South Korea. Air was so toxic that several schools moved or closed down out of concern for students' health. Air pollution emitters in Ulsan had to pay large amounts of compensation to nearby farmers every year for losses caused by sulfur dioxide (SO_2) and fluorine gas emissions.[35]

Part of the solution to the problems posed by Korea's industrial complexes was the national EIP. There were other efforts, including the "energy block platforms," which targeted industrial complexes. The aim of these platforms was to develop a solution for blocks of industrial scale, independent and self-sufficient energy grids, which could be targeted at both domestic and export markets. The Korea Micro Energy Grid consortium led by Samsung C&T was sponsored by MKE/MOTIE from 2011–2013 with a budget of US$100 million.[36]

Jeju Smart Grid project and remote island microgrids

About 300 of the approximately 3,000 islands surrounding the southern half of the Korean Peninsula are inhabited and the government owned Korea Electric Power Corporation (KEPCO) has active microgrid and smart grid projects underway on a number these islands. By far the largest of these projects is well underway on the island province of Jeju, one of the largest such undertakings in the world.

In 2009, the year after President Lee Myung Bak announced low-carbon green growth as the national vision, Korea developed a proactive and ambitious plan to build a smart grid testbed on Jeju Island. The project was the largest one in Korea and involved hundreds of foreign and domestic companies. Korea's national government, KEPCO, the Korea Smart Grid Institute (KGSI), the Jeju Special Autonomous Province and the Korea Smart Grid Association were key players in the Jeju Island pilot project. The entire initiative involved a number of Korea's global corporations and hundreds of companies. These companies and organizations, along with academic and government institutions, were to test technologies and business models that could ultimately roll out in selected cities on the Korean mainland.[37]

The first and best known of Korea's small island microgrid projects is the one on tiny Gasa Island, located off the southwestern coast of Korea, not far from the much larger Jeju Island. The nation's first completely automated energy management system, developed by KEPCO, powers Gasa Island's homes and small businesses. In case of outages, the fully charged batteries in its electricity storage system (ESS) can provide enough electricity to supply the entire island for up to 24 hours. Residents of Gasa Island have expressed satisfaction with the microgrid facilities. A civic leader on the island noted that

> In the past, fishermen here have had to export our seaweed for cheap because we didn't have enough power to sustain a drying facility. We've had to pay other fishermen on mainland Jindo to dry and package our seaweed and fish to sell to our Japanese buyers. Thanks to the microgrid facility, we have a sustainable source of electricity."[38]

As of 2015, the Gasa Island project provided 80 percent of the island's total electricity, and the government had worked with KEPCO to transfer the system to 86 other islands around the peninsula.

As this is written, the Jeju project has already made significant progress as noted by the governor of the Jeju self-governing province in a 2018 speech. The overall goal as of 2018 was to make Jeju a carbon-free island by 2030 and the following milestones indicate some of the progress:

- In 2014, the renewable energy supply stood at 6.4 percent, but it had increased to 13.6 percent by 2018.
- In September 2017, Jeju started to operate the first commercial offshore wind farms in Korea.
- In March of 2018 Jeju entered "the era of 10 thousand EVs" and half of the total electric vehicles registered nationally were in Jeju.[39]

Based on its experience and success at home, Korea promoted island microgrids as a promising export. As of 2015, the Korean government had signed MOUs with Kuwait and Qatar and countries like Colombia, Peru and Chile.[40]

In August 2018 the state-invested Korea Institute for Advancement of Technology (KIAT) signed an MOU with Ecuador's electricity and renewable energy ministry to cooperate in setting up a microgrid system in the Galapagos archipelago. The project, worth US$6.2 million, was funded by Korea's trade ministry with its official development assistance (ODA) budget. The Galapagos Islands, which consist of islands and rocks formed at least 5 million years ago, are considered one of the most bio-diverse places on earth, with great scientific significance.[41]

Application of blockchain technology

In 2018 KEPCO announced that it would integrate blockchain technology with its existing microgrid networks to develop the KEPCO Open Microgrid Project. Its existing microgrid consisted of photovoltaics, wind turbines and energy storage devices. While such microgrids were functional, they struggled to supply stable power. The new system would ensure that there was no waste of energy by using power to gas (P2G) conversion in which leftover energy is converted to hydrogen, which allows it to be stored and re-converted to electricity when the need arises. In that manner, the Open Microgrid Project embraced the goal of President Moon Jae-In's government to create a hydrogen-based economy. The use of blockchain would ensure improved recording of energy generation and conversion figures, allowing users to easily buy and sell energy from and to the grid.[42] KEPCO's long-term aim was to create the world's first mega-wattage energy-independent microgrid.

A progress report

A 2017 performance review by the OECD noted that Korea needs to put its green growth vision into action. It noted that, Korea's energy mix was dominated by fossil fuels. As shown in Figure 7.2, Korea's greenhouse gas emissions rose by 39 percent between 2000 and 2013, the second highest growth rate in the OECD, after Turkey.[43]

The report noted that Korea is the most densely populated country in the OECD. Consequently, it is one of the countries most affected by fine particle air pollution, a particular concern in the national capital metropolitan area around Seoul, which is home to half of Korea's population. The problem, the OECD observed, was exacerbated by fine particle pollution that drifted over the Korean Peninsula from China.

Korea's lack of progress in reducing GHGE as of 2018 does not mean a total lack of progress. For evidence of this, one need only look to the passage of laws, the drafting of five-year and long-term policies and the establishment of many institutions with responsibility to address the problem. The 2009 Framework Act on Low Carbon Green Growth was the first such law in the world and ensured that green growth would be approached in a comprehensive and systematic manner.[44]

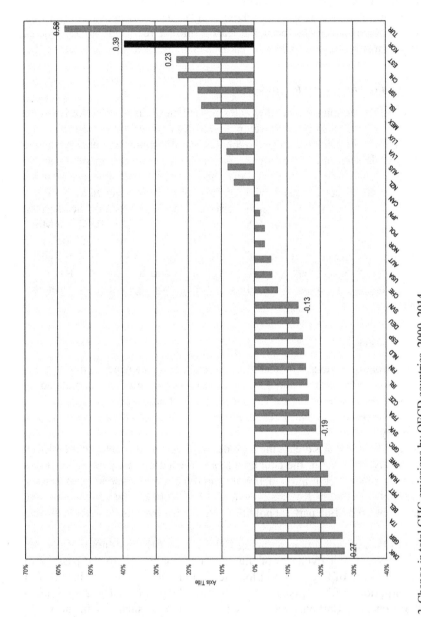

Figure 7.2 Change in total GHG emissions by OECD countries, 2000–2014

Source: OECD Environment Statistics Database 2016.

Retrospect and prospect: future electricity networks

In retrospect, there is a strong link between the origins of Korea's digital development in the 1980s and the challenge it faces to reshape the electric power grid in
Korea into a suitable smart and sustainable network or grid. As noted in the first
edition of this book, as early as 1980, KEPCO had installed fiber throughout its
network, gambling that, once it had the infrastructure, the government would be
forced to allow KEPCO to use it more productively. This was interesting, given
that KEPCO was fully government-owned and under the jurisdiction of the Ministry of Commerce and Industry (MCI). At the same time, KT had created its own
fiber networks with government support, partly financial. A dispute ensued when
both KEPCO and KT applied for a license to lease fiber to telecom carriers. Purely
from a capacity standpoint, KT's infrastructure was sufficient, while KEPCO's
was redundant. This situation pitted the MCI against the Ministry of Information
and Communications, but they could not engage in a public battle because that
would have revealed that MCI had allowed KEPCO to take matters into its own
hands with taxpayer's money.[45]

The history makes a powerful point. KEPCO needed a national fiber optic network in the early 1980s in order to efficiently administer the national electric
power grid. Today, as this is written, Korea's status as a world leader in digital
network infrastructure gives it a big edge over other countries in the development
of smart grids, which are based on the convergence of ICT with electric power
grids. In short, despite Korea's lack of significant progress to date in reducing
greenhouse gas emissions, this chapter suggests that it has embarked on another
transformation that will be perhaps as significant as the telecommunications revolution of the 1980s. Both transformations center on digital technologies and can
be aptly characterized as network-centric development. The outcome of this transition to green growth will be ultimately significant not only for South Korea but
for all nations and peoples on planet earth.

Notes

1 Belkhir, Lotfi and Ahmed Elmeligi. (2018). Assessing ICT global emissions footprint:
  Trends to 2040 & recommendations. *Journal of Cleaner Production*, 177, pp. 448–463.
2 United Nations. (1987). *Report of the World Commission on Environment and Development: Our common future*. United Nations.
3 Souter, D. (2012). *ICTs, the internet and sustainability: A discussion paper*. Winnipeg:
  International Institute for Sustainable Development, p. 1.
4 Souter, D. (2012). *ICTs, the internet and sustainability: A discussion paper*. Winnipeg:
  International institute for sustainable development, p. 2.
5 Souter, D. (2012). *ICTs, the internet and sustainability: A discussion paper*. Winnipeg:
  International institute for sustainable development, p. 4.
6 Souter, D. (2012). *ICTs, the internet and sustainability: A discussion paper*. Winnipeg:
  International institute for sustainable development, p. 3.
7 Souter, D. (2012). *ICTs, the internet and sustainability: A discussion paper*. Winnipeg:
  International institute for sustainable development, p. 6.
8 Souter, D. (2012). *ICTs, the internet and sustainability: A discussion paper*. Winnipeg:
  International institute for sustainable development, p. 7.

 9 Madden, P. A. (2008). *Connected: ICT and sustainable development*. London: Forum for the Future: Action for a Sustainable World, p. 6.
10 Madden, P. A. (2008). *Connected: ICT and sustainable development*. London: Forum for the Future: Action for a Sustainable World, p. 8.
11 Madden, P. A. (2008). *Connected: ICT and sustainable development*. London: Forum for the Future: Action for a Sustainable World, p. 9.
12 Madden, P. A. (2008). *Connected: ICT and sustainable development*. London: Forum for the Future: Action for a Sustainable World, p. 12.
13 Earth Institute at Columbia University, Ericsson. (n.d.). *ICT and SDGs: How information and communications technology can achieve the sustainable development goals.* Earth Institute at Columbia University and Ericsson, p. 2.
14 Earth Institute at Columbia University, Ericsson. (n.d.). *ICT and SDGs: How information and communications technology can achieve the sustainable development goals.* Earth Institute at Columbia University and Ericsson, p. 2.
15 Catholic Relief Services. SDG ICT playbook: From innovation to impact. Retrieved from December 20, 2018, from www.crs.org/our-work-overseas/research-publications/sdg-ict-playbook
16 Souter, David, Don MacLean, Ben Akoh and Heather Creech. (2010). *ICTs, the internet and sustainable development: Towards a new paradigm.* Winnipeg, Canada: International Institute for Sustainable Development, p. 4.
17 Young, S. (2016, November 4). What are you going to teach for global citizenship education? Seoul, Korea. Unpublished document.
18 Kim, Sung-Young and Elizabeth Thurbon. (2015). Developmental environmentalism: Explaining South Korea's ambitious pursuit of green growth. *Politics and Society*, Sage Publications, p. 2.
19 Kim, Sung-Young and Elizabeth Thurbon. (2015). Developmental environmentalism: Explaining South Korea's ambitious pursuit of green growth. *Politics and Society*, Sage Publications, pp. 2–3.
20 Kim, Sung-Young and Elizabeth Thurbon. (2015). Developmental environmentalism: Explaining South Korea's ambitious pursuit of green growth. *Politics and Society*, Sage Publications, pp. 10–11.
21 Interview with Dr. Young Soogil, November 27, 2018, co-chair of the Presidential Committee on Green Growth under President Lee Myung Bak, interviewer James F. Larson.
22 Interview with Dr. Young Soogil, November 27, 2018, co-chair of the Presidential Committee on Green Growth under President Lee Myung Bak, interviewer James F. Larson.
23 Global Green Growth Institute. (2015). *Korea's green growth experience: Process, outcomes and lessons learned.* Seoul: Global Green Growth Institute, p. 1.
24 Interview with Dr. Young Soogil, November 27, 2018, co-chair of the Presidential Committee on Green Growth under President Lee Myung Bak, interviewer James F. Larson.
25 Interview with Dr. Young Soogil, November 27, 2018, co-chair of the Presidential Committee on Green Growth under President Lee Myung Bak, interviewer James F. Larson.
26 Kim, Jaewon. (2017, May 28). South Korea's new energy plan puts pressure on Kepco. *Nikkei Asian Review*. Retrieved November 21, 2018, from https://asia.nikkei.com/Business/South-Korea-s-new-energy-plan-puts-pressure-on-Kepco
27 Korea Power Exchange (KPX). The Korea power exchange and electricity market. Brochure Retrieved November 22, 2018, from www.kpx.or.kr/www/download BbsFile.do?atchmnflNo=16564
28 Kim, Sung-Young and John A. Matthews. (2016, December 15). Korea's greening strategy: The role of smart microgrids. *The Asia-Pacific Journal Japan Focus*, 14(24), p. 6. Retrieved November 21, 2018, from https://apjjf.org/2016/24/Kim.html

29 Kim, Sung-Young. (2018). Hybridized industrial ecosystems and the makings of a new developmental infrastructure in East Asia's green energy sector. *Review of International Political Economy*. doi:10.1080/09692290.2018.1554540.

30 Kim, Sung-Young. (2018). Hybridized industrial ecosystems and the makings of a new developmental infrastructure in East Asia's green energy sector. *Review of International Political Economy*, p. 8. doi:10.1080/09692290.2018.1554540.

31 Bulkin, Bernie. (2014, May 2). South Koreans are leading the way in their vision for 'smart grid'. *The Guardian*. Retrieved November 20, 2018, from www.theguardian.com/sustainable-business/smart-grid-south-korea-rethinking-electricity

32 Kim, Sung Young and John A. Matthews. (2017, March 29). Korea's strategy for smart grids. *The Korea Times*. Retrieved November 20, 2018, from www.koreatimes.co.kr/www/opinion/2017/03/162_226606.html

33 Whang, Jaehong and Woohyun Hwang, Yeuntae Yoo and Gilsoo Jang. (2018, September 30). Introduction of smart grid station configuration and application in Guri branch office of KEPCO. *Sustainability*, MDPI.

34 Kim, Eunice Jieun. (2017, June). *Case study: Greening industrial parks – a case study on South Korea's eco-industrial park program*. Seoul: Global Green Growth Institute.

35 Kim, Eunice Jieun. (2017, June). *Case study: Greening industrial parks – a case study on South Korea's eco-industrial park program*. Seoul: Global Green Growth Institute, pp. 5–6.

36 Kim, Sung-Young and John A. Matthews. (2016, December 15). Korea's greening strategy: The role of smart microgrids. *The Asia-Pacific Journal Japan Focus*, 14(24), p. 6. Retrieved November 21, 2018, from https://apjjf.org/2016/24/Kim.html

37 *South Korea: Smart grid revolution*. Special report by Zpryme's smart grid insights, Zpryme Research and Consulting, July 2011.

38 Kim, Ji-Yoon. (2015, June 3). Global sustainable energy starts on Korea's islands. *Korea Joongang Daily*. Retrieved November 21, 2018, from http://koreajoongang daily.joins.com/news/article/article.aspx?aid=3004894

39 Won, Hee-ryoung, Governor of Jeju province, speech delivered March 13, 2018. Retrieved November 22, 2018, from www.unosd.org/content/documents/3861Keynote%20Speech%20Governor%20Won.pdf

40 Kim, Ji-Yoon. (2015, June 3). Global sustainable energy starts on Korea's islands. *Korea Joongang Daily*. Retrieved November 21, 2018, from http://koreajoongang daily.joins.com/news/article/article.aspx?aid=3004894

41 Ko, Jae-man and Cho Jeehyun. (2018, August 24). Korea to set up microgrid system in a Galapagos Island. *Pulse*. Retrieved November 21, 2018, from https://pulsenews.co.kr/view.php?year=2018&no=532489

42 Uti, Tokoni. (2018, November 20). Korean KEPCO to use blockchain in energy micro grid. *Blockchainreporter*, Retrieved November 22, 2018, from https://blockchainreporter.net/2018/11/20/korean-kepco-to-use-blockchain-in-energy-micro-grid/

43 OECD, *OECD environmental performance reviews: Korea 2017*. Retrieved November 22, 2018, from www.oecd.org/korea/oecd-environmental-performance-reviews-korea-2017-9789264268265-en.htm

44 Retrieved from http://17greengrowth.pa.go.kr/english/

45 Kushida, Kenji and Seung-Youn Oh. (2007). The political economies of broadband development in Korea and Japan. *Asian Survey*, 47(3), p. 494.

8 The Olympics and digital development

"Our Korean forefathers failed to ride the surging tide of the Industrial Revolution, consequently failing to modernize, remaining poor and backwards. We've made progress since then, but another revolution is before us and if we fail to adapt ourselves to the changes, future generations will inherit poverty and backwardness just as we did a generation ago."

Statement by Minister of Communications Oh Myung, on the eve of the 1988 Seoul Olympics

In historical perspective the 1988 Summer and 2018 Winter Olympics effectively frame Korea's remarkable and rapid digital development. This chapter examines the relationship between the Olympics and digital development at those two points in time. In 1988 the global system of industrial mass media that was dominant for over a century was still strong, but the transition to a networked information economy was beginning, most especially in Korea with its "1980s telecommunications revolution." Thanks in part to this early start in digital network infrastructure development, by the time Korea hosted the Pyeongchang 2018 Winter Olympics the nation was an acknowledged world leader in broadband network infrastructure. Accordingly, the 2018 Winter Games were planned to showcase Korea's digital prowess, with emphasis on next generation networks for both commercial 5G and public safety purposes.

The Olympics bear directly on Korea's digital development and the modern transformation of the Olympics in three major ways. First, the infrastructure and preparations needed to host the Olympics in both cases helped to shape Korea's approach to building its own network infrastructure. At the time that the 1988 Olympics were awarded to Seoul, Korea did not yet have the infrastructure in place to host such a global media event. As Pyeongchang approached, international standards for 5G were not yet finalized, yet Korea's electronics companies and telecommunications service providers, along with their international partners, were hopeful that the 5G standards on display in the 2018 Winter Olympics would receive international (3GPP) approval.

Second, the Olympic Games gave Korea a platform to display its broadcasting and telecommunications prowess for the whole world to see. In 1988, before the internet, World Wide Web and social media, this was a coming out party of sorts for a nation that had been cut off or obscured to much of the world during the long Cold War. Its broadcasting, telecommunications and electronics capabilities were a big part of this story. The 2018 Pyeongchang Winter Olympics were planned to offer visitors and the world a glimpse of what next generation networks and the future of digital communications would look like.

Third, the experience of Pyeongchang in 2018, followed by Tokyo in 2020 and the Beijing Winter Olympics in 2022 seem certain to display both the influence of digital technologies on Olympics sport and the manner in which the Olympics may influence next generation networks. Industry watchers are already asking how the Olympics will shape 5G. A 2016 report noted that the determination of East Asian nations to showcase 5G in the 2018 and 2020 Olympics seems to be "accelerating the global development of the technology."[1] Korea Telecom and other Korean mobile service providers anticipated that the 5G technologies used in the Pyeongchang Olympic Games would be compatible with the first phase of the 5G standard, scheduled to be finalized by the international standards body 3GPP in 2018.

Once 5G becomes commercially available, there is little doubt that it may transform the experience of watching live sports. For example, spectators may be able to watch an event from different vantage points, switching between cameras mounted on an athlete's helmet to a bird's-eye view to a conventional side-on viewpoint. Next generation networks promise to allow spectators to watch 3D holograms or 360-degree views of the sports action using virtual reality headsets. Augmented reality may also allow spectators to focus on particular athletes and retrieve information about them.[2]

Lee Byeong Moo of KT left little doubt about the company's intentions. "Presenting 5G service for the first time in the world throughout PyeongChang 2018 will be a triggering point for Korea to lead the 5G industry, which aligns with KT's goal," he said. "As the Olympics is a world event, being the official telecom partner will be our starting point for 5G."[3]

This chapter addresses the following questions:

- How does the use of new digital media change the Olympic experience for athletes, fans, coaches and other members of the Olympic family?
- How do the communications requirements of the Olympic Games influence digital innovation and next generation networks?
- To what extent and with what impact will the communication networks for the 2018 Winter Olympics help to promote the nation's role as a next generation "testbed for the world"?
- How does the presence of cutting-edge digital networks change the role of the Olympics as a major media event?

The Olympics in the industrial mass media era

Summarizing an influential strain of 20th century communications scholarship, Katz and Dayan defined media events as the "high holidays of mass communication." Their book focused on historic occasions such as the Olympic Games,

Anwar el-Sadat's journey to Jerusalem and the funeral of John F. Kennedy. Most such events involved states, were televised live and constituted a new narrative genre. Media events, they noted, used the power of electronic media to command global and simultaneous attention.[4]

Audiences

During the mass media era, the Olympic Games were consistently ranked as one of the largest global media events, and the Seoul Olympics were no exception. This pattern was discernable despite the wide variation across countries and companies in measurement methods, as noted by a major study of television in the Olympics conducted around the 1992 Barcelona games.[5] The audience for the opening ceremony telecast from Seoul was estimated at over 1 billion people around the globe, nearly double that of the 522 million viewers of the opening ceremony for the Los Angeles Olympics in 1984.[6] By most measures, the 1988 Seoul Olympics were the largest planned television event in history to that date.

During September of 1988, people around the world focused their attention through television on South Korea and its capital city of Seoul. Not since the Korean War had there been such massive media attention. Indeed, as noted in the opening chapter, the strong tendency of mainstream media had been to pay attention to Korea only sporadically, in times of crisis, a pattern that had persisted through the long Cold War period. In addition to the massive focus of global attention in 1988, the summer games opened a window on the world for citizens of a nation formerly known as the "Hermit Kingdom." In a first for Olympic television, the presence of the Armed Forces Korea Network (AFKN), a broadcasting service for U.S. military in Korea made it possible for Koreans to view the full Olympic coverage another nation's broadcaster. AFKN broadcast the coverage provided by NBC, the Olympic broadcast rights holder from the United States.[7]

The large worldwide viewership for the Seoul Olympics illustrated a 20th century trend toward increased public reliance on television versus print and other media for entertainment and information. However, even at that time it was apparent that such large global audiences were more than simply a response to new media technologies. The large viewing audiences persisted "in the face of a widely acknowledged trend toward more fragmented and specialized audiences, facilitated by the appearance of new technologies for gathering, storing, and disseminating information in many forms, including video."[8]

However, global television alone cannot explain the full significance of the Seoul Olympics for Korea's socioeconomic development and the role of digital networks in it. The larger import of the 1988 Seoul Olympics only becomes apparent in relation to four transformations that occurred internationally, in Korea and in the Olympic movement during the 1980s. The first of these is the digital network revolution that is a central focus of this book. By the 1980s, a modern telecommunications infrastructure to support the administration and global broadcast of the Games was one prerequisite for host cities. While Korea did not possess such an infrastructure in 1981 when the games were awarded to Seoul, the prospect of hosting the 1988 Olympics gave extra impetus to the nation's tele-communications revolution of the 1980s. A second major transformation involved geopolitics.

The Olympics, geopolitics and democratization

In geopolitical terms, the Seoul Olympics contributed to the end of the Cold War in Asia. From the very first discussions in 1978 of a possible bid to host the Olympics, Korean leaders saw the Games as an instrument of foreign policy that might address the central political problem of national division. The Park Chung Hee administration also thought that Korea had reached a turning point because of high economic growth and might be able to emulate Japan's earlier experience with the Olympics. President Park also believed that staging the Olympics would give Korea an opportunity to join the ranks of the advanced nations. In the actual event, despite intensive negotiations by Seoul and the International Olympic Committee (IOC) with North Korea between 1984 and 1987, there was no agreement to co-host the Olympics, and North Korea did not participate. However, the 1988 Olympics became an ideal vehicle for South Korea's "Northern Policy," which aimed to improve relations between North and South Korea by opening up friendly relations with countries that had been completely cut off during the long Cold War. These included the Soviet Union, the socialist nations of Eastern Europe, China and Vietnam.[9] For evidence of the economic impact of the successful Northern Policy and the end of the Cold War, one need only look at Korea's exports of semiconductors, mobile handsets, displays and television sets to those former socialist-bloc nations with which there had been no contact.

A third major transformation during the 1980s was the move toward political liberalization and democracy in South Korea. The 1988 Olympics were awarded to Seoul less than two years after the political turmoil of 1980. This timing virtually ensured that public pressures for democratization and free elections would play a role leading up to the Seoul Olympics. Indeed, by June of 1987 protestors against the government of President Chun Doo Hwan were demanding elections, greater press freedom and other reforms.

On June 29, 1987, Roh Tae Woo, President Chun's close associate and former head of the Seoul Olympic Organizing Committee, delivered a nationally televised speech in which he accepted all of the major opposition demands point by point including direct presidential elections and amnesty for opposition leader Kim Dae Jung. That speech marked a decisive turning point toward political liberalization and democracy in Korea.

Olympic sponsorship and Samsung's growth as a global corporation

The fourth transformation in the 1980s took place within the Olympic movement itself. Under the leadership of Juan Antonio Samaranch, the IOC focused its attention on the questions of how to ensure both universal participation and stable financing. The introduction of The Olympic Program (TOP) to coordinate global corporate sponsorship gave the IOC a second major source of income to supplement television rights revenues.

> The Seoul Olympics, together with the Calgary Winter Games earlier in 1988, ushered in a new era of commercial involvement with the Olympics. For the first time in Olympic history corporate sponsorship activities were planned and carried out on a truly global scale. . . . The initiative was called The Olympic Program (TOP), and it was negotiated over a period of years by the International Olympic Committee, the organizing committees for the Seoul and Calgary Games, and more than 150 separate national Olympic committees.[10]

Furthermore, TOP marked a new stage in the interdependent development of both television broadcasting and the marketing practices of transnational corporations.

According to most industry analyses the key factor behind the worldwide growth of sponsorship is its particular suitability as a global communications medium. This appealed to Juan Antonio Samaranch, president of the IOC, who had two main motivations for creating the TOP program. First, he felt that the IOC was becoming too dependent on television rights as a source of revenue. Second, he saw the Olympics as being dominated by teams with superior financial resources. Both of these problems could be addressed by TOP.[11]

There were nine TOP sponsors for the 1988 Seoul Olympics, including Coca-Cola, Kodak, Sports Illustrated/Time, VISA, Brother, Philips, 3M, Federal Express and Matsushita Electric (Panasonic). Notably no Korean companies were in this elite group of sponsors.

Samsung's involvement with the Olympic Games began as a local sponsor of the 1988 Seoul Olympics. Along with Goldstar, Philips from the Netherlands and Matsushita of Japan it was named an official supplier of electronics products for the Seoul Games. The majority of suppliers for the Seoul Olympics were Korean companies, including Hyundai, which introduced the Stellar 88, official car of the Games.[12]

> In a distinctive effort by a supplier, Samsung Electronics gave each of the 100,000 persons attending the opening ceremony a small (4.5cm × 2cm × 1cm), clip-on FM receiver weighing just a few ounces and earphones through which the spectators could tune to a narrative description of the ceremony in English, German, French, Korean, Russian, Spanish, Arabic, or Japanese. This program had several advantages: It was the first time in Olympic history that such technology was used, it targeted a highly influential group of spectators, and it generated a great deal of additional television and press coverage.[13]

When all was said and done, sponsorship activities for the Seoul Olympics brought in US$140 to US$150 million, compared with more than US$407 million generated by television rights. Assuming limited revenues from licensing and ticket sales, sponsorship accounted for something more than one quarter of Olympic revenues. That proportion would change dramatically over the years leading up to the 2018 Pyeongchang Winter Olympics. Also, the efforts by Samsung as a supplier would foreshadow Samsung's future role as a TOP sponsor.

On a corporate level, Samsung Electronics today is the most prominent symbol of South Korea's network-centric digital development. Among the nation's large conglomerates, it became the leading manufacturer of semiconductors, television sets and more recently smartphones. Beginning with the Nagano Winter Olympics in 1998, Samsung joined TOP as a worldwide Olympic partner in the wireless communication equipment category, providing both mobile phones and its own proprietary wireless communication platform called Wireless Olympic Works (WOW). In August of 2014 Samsung extended its TOP sponsorship deal with the IOC through 2020, ensuring that it would be the sponsor during the Pyeongchang 2018 Winter Olympics.[14]

The International Olympic Committee (IOC) and the organizations within the Olympic movement are entirely privately funded. As of 2018, 47 percent of Olympic marketing revenues come from the sale of broadcast rights and 45 percent from TOP sponsorship activities for a total of 92 percent of all income. Five percent of revenues came from ticketing and 3 percent from licensing.[15] Given the close relationship of global television reach and the value to corporations of sponsorship activities, this is strong empirical evidence of the central role that networks, both broadcast and digital, have come to play in the modern Olympics.

Telecommunications and the Olympics[16]

Although Korea's telecommunications revolution of the 1980s culminated in the highly successful 1988 Seoul Olympics, it is well to remember that the Seoul Games occurred in the pre-internet era, four years before the launch of the World Wide Web and over a decade before the arrival of social media and smartphones. Nevertheless, the Seoul Olympics set a new high water mark for both the number of broadcast personnel and the size of the technical infrastructure involved. It is worth recounting the role of telecommunications in the Seoul Olympics, keeping in mind that progress in digital development, like other areas of science and technology, is cumulative and subject to the "on the shoulders of giants" effect long noted by economists.

The award of the 1988 Olympics to Seoul marked a major departure for the modern Olympic movement for two reasons. One was the uncertainties on the divided Korean Peninsula, with growing public opposition to the Chun Doo Hwan government. The second was that South Korea was still widely considered to be a developing country. Prior to Seoul, the IOC had awarded the Olympics only to cities in industrialized nations. Korea's status in 1981 significantly magnified the challenge it faced to successfully host the Olympic Games.

To address the Olympic project, South Korea assembled a large group of technocrats, many of them educated in technical disciplines in the U.S. During the Seoul Olympics project, many of these individuals from industry, government and academia were transferred temporarily to the Seoul Olympic Organizing Committee, host broadcaster KBS/SORTO or related organizations. In preparing for the telecommunications infrastructure required for the Olympics, five major parties were involved, as follows:

1 Private Branch Exchange (PBX) – city of Seoul
2 The Korean Broadcasting System (both KBS and KBS/SORTO)
3 The Seoul Olympics Organizing Committee
4 The Korea Telecommunications Authority
5 Press organizations (print media)

The nation's leading technocrats forcefully seized upon both the Asian Games in 1986 and the 1988 Olympics as vehicles to speed the nation's development of high technology. On the eve of the Seoul Olympics, Dr. Oh Myung, then serving as Minister of Communications declared,

> Our Korean forefathers failed to ride the surging tide of the Industrial Revolution, consequently failing to modernize, remaining poor and backwards. We've made progress since then, but another revolution is before us and if we fail to adapt ourselves to the changes, future generations will inherit poverty and backwardness just as we did a generation ago. [17]

During the decade leading up to the Seoul Olympics, Korea experienced one of the most rapid growth rates in telecommunications of any nation in the world. As described earlier in this book, it went from having a desperate telephone service backlog in 1980 to completion of a nationwide public switched telephone network in June of 1987. This rapid development had important implications for both the administration of the Olympics and television and media coverage of the Games. Korea's basic telecommunications network was state of the art. This meant, among other things, that it could innovate. Dr. Oh Myung, as Minister of Communications, announced that Korea was ready to show the world that it could provide such advanced systems as hand phones, trunking radio systems, Group 4 fax and color graphic systems.

The technical infrastructure built for the international telecast of the Seoul Olympics was massive. It had to serve the needs of more than 10,000 accredited broadcast personnel. Moreover, the Olympic competition covered by the host broadcaster KBS/SORTO involved 25 sports in 34 competition sites in Seoul, Busan and other Korean cities as well as opening and closing ceremonies and other official ceremonies within the scope of the Games.

Responsibility for providing and operating all telecommunications circuits required to broadcast the Seoul Olympics fell to the Korea Telecommunications Authority, working with Intelsat internationally. Owing to the near-universal participation in the Seoul Olympics by countries around the world, demand for satellite links exceeded Korea's capacity. Intelsat eventually used a total of nine satellites to broadcast the Seoul Olympic events – four in the Atlantic Ocean region, three in the Indian Ocean region and two in the Pacific Ocean region. Demand for television transmission of the Summer Games was so great that a near-saturation level was reached on the Indian and Pacific Ocean television channels. As the Games approached, KTA also leased a portable K-band antenna from Teleglobe Canada and during the Games called in IDB Communications, a Los Angeles-based transmission service to fly two KU-band earth stations to Seoul for use in radio transmissions.

The growth in size of the Olympics involved dramatic increases in the number of broadcasting, press and support personnel and mounting demands for the rapid dissemination of games results and other Olympic information to media representatives. This development was partly due to the more general computerization of television news, sports and other operations that was taking place around the world. When Seoul was awarded the Olympics in 1981, one of the first questions asked was whether Korea would be able to develop satisfactory information systems. At that early date, even the nation's basic telephone system was suffering from a growing backlog in service and the country's electronics and computer industries were in embryonic stages.

In these circumstances, the government resolved to use the Olympics as an opportunity to foster Korea's information industries and it set three general goals. First, it would carry out preparations for the Olympics in conjunction with its Telecommunications Modernization Plan. Second, it would provide state-of-the-art information services for the Games. Finally, it would seek to spread the benefits of the Olympics project across all relevant industries and users.

In February 1984 the Seoul Olympic Organizing Committee decided to develop its own systems instead of adopting the system that had been used in prior Olympics. Teams of experts were dispatched to cities that had hosted recent Olympics to survey the options. Upon returning from Los Angeles in June 1984, a group led by Vice Minister of Communications Oh Myung issued a report that established the direction of Korea's planning for telecommunications and information systems. Its key recommendations were:

1 To expand the information and telecommunications groups in the Ministry of Communications, Korea Telecom and the Seoul Olympic Organizing Committee;
2 To expand facilities for international television broadcasting, including additional earth stations and portable earth stations;
3 To provide new services that drew attention in Los Angeles, including an electronic messaging system, electronic mailbox and card telephones;
4 To integrate computer and communications systems;
5 To foster domestic manufacture of telecommunications equipment for the Olympics; and
6 To set up a fully computerized support system for management of the Games.

Ultimately, four different organizations developed computer systems that were integral to the Seoul Olympics. The two major systems were the Wide Information

Network System (WINS), developed by DACOM and tested in an earlier more limited version during the Asian Games, and the Games Information Online Network (GIONS). The WINS system provided electronic mail and information retrieval services in four languages: Korean, Spanish, English and French. WINS terminals were widely available in the International Broadcast Center, at Olympic venues and internationally through packet switched and telex networks. Much of the Games information came through a link with GIONS, which transferred results through WINS as soon as they were available.

WINS and GIONS set three new standards for Olympic information systems. First, they were the first full-scale integration of computer and communications systems, making both the information retrieval service and e-mail available to domestic and overseas users in more than 70 countries. Second, GIONS was built around distributed processing architecture, with a pair of microcomputers at each Olympic venue, so that if one unit failed the other could serve as a backup, and the venue systems were designed to operate independently should the host system fail. Third, there was a direct interface between the information systems and timing or measurement devices for seven sports, including swimming, cycling, shooting, archery, gymnastics and some track and field events. A report by the Korea Development Institute singled out the development of the WINS system by a domestic technology team as a significant step forward to establish Korea as an information society.

The Seoul Olympics gave impetus to the development of several broadcast specific technologies, including graphics, character generators known as PRISM units and the inauguration of the first Korean teletext service using equipment developed by the KBS technical laboratories. Korean manufacturers provided KBS with an estimated 45 percent of all electronic equipment used at the Games.

The Olympics in the digital network era

Three decades after the Seoul Olympics Korea hosted the 2018 Pyeongchang Winter Olympics. By that time, the advent of the World Wide Web, broadband internet and near-universal use of mobile phones had transformed the Olympics as a global media event. The Olympic Games are now a "bring your own device," multiple screen, social media event suitable for the early decades of the hyperconnected digital network era.

Korea Telecom, the official communications service provider for Pyeongchang, publicly declared its intent to provide the world's first 5G Olympics. Not to be outdone, mobile operators for the 2020 Tokyo Summer Olympics also announced plans to use 5G to provide innovative services to spectators, viewers and organizers.[18] Given Korea's current status as a world leader in mobile broadband networks, expectations for 2018, both within Korea and globally, were extremely high.

Audiences and the multi-screen experience

The Pyeongchang Winter Olympics set a new high water mark for the overall size of the global viewing audience, but with relatively more people turning to digital and cable channels than to broadcast television. The IOC estimated that 28 percent of the world's population watched Pyeongchang 2018 broadcast coverage. It also reported that digital viewership was more than double that of Sochi 2014 with 3.2 billion video views and 16.2 billion minutes viewed. All in all, the IOC reported that the Pyeongchang games were the biggest Olympic Winter Games ever on social media platforms.[19]

Beginning with the London 2012 Summer Olympics, audiences began using multiple devices to view the Games. In London the Olympics presented fans, athletes, coaches, judges and members of the Olympic family with the option of viewing on smartphones or tablets rather than a television set and also the possibility of sharing photos and video via social media. Google conducted research in the UK on Olympic-related media consumption across four screens: television, smartphones, tablets and computers. The key findings were as follows:

- One in three people followed the Olympics on multiple screens on any given day.
- Smartphones extend event engagement, both in and out of the home.
- The Olympics caused people to try new things, in particular on phones or tablets.
- Fifty percent of spectators contributed to the Olympics digital afterlife by sharing photos and videos online.[20]

The Google study showed that television was still the high-reach medium, followed in rank order by computers, smartphones and tablets. One in three Olympics followers used more than one screen in a day, with the number of screens correlating to the expressed excitement of the follower about the Olympics. Those who were more excited or interested also spent more minutes per day following the Olympics.

The Google research also showed three different types of multi-tasking. One involved following the same event on the phone and with another screen. Another was following two events, one on the phone, and the other on a different screen. Finally, some followers followed the Olympics on their phones while doing something unrelated.

Nielsen reported that the Beijing Olympics in 2008 drew the largest ever global TV audience. According to the Nielsen measure of reach, 4.7 billion people tuned into the Beijing Olympics amounting to more than two out of three people worldwide. This exceeded the 3.9 billion who followed the Athens Olympics in 2004 and the 3.6 billion who watched the Sydney Games in 2000.[21] Estimates of the global reach of the London 2012 Olympics were in the same ballpark.

As noted on the official IOC website, "Broadcast coverage is the principal means for people around the world to experience the magic of the games."[22] That statement reflects the traditional role of television in the industrial mass media era as the principal driver of funding, global popularity, representation and promotion of the Olympic Movement, the Olympic Games and Olympic values. However, that role is changing because of the intrusion of digital networks and new media into the Olympic viewing experience. The growing importance of digital media relative to traditional television broadcasting led NBC in the United States to introduce a new metric called total audience delivery (TAD). TAD accounts for expected viewership from TV, as well as video-on-demand (VOD), over the top (OTT) and mobile platforms. NBC based its advertising sales for Pyeongchang entirely on expected TAD.[23]

One way in which the Olympic movement measures its worldwide television broadcasting activities is by total hours of coverage. This includes free-to-air coverage, cable and satellite TV and more recently digital video received via websites, mobile sites or apps.

By the time Beijing hosted the Summer Olympics in 2008, large quantities of online video content were available for the first time in Games history. The IOC reported that from a sample of sites for which statistics were available there were a total of 8.2 billion page views and over 628 million video streams. The IOC also launched its own internet channel, "Beijing 2008" through the You-Tube platform to make video highlights from the Games available in territories where digital video-on-demand rights had not been sold. The IOC report on Beijing 2008 concluded that the broadcast and consumption through multi-platforms in all territories established two things. First, it had become realistic to broadcast and consume each and every moment of the Games, not just on-demand, but live (also referred to as linear television). Second, the 2008 Games demonstrated that multi-platform offerings through TV, the internet and on mobile devices, far from cannibalizing TV ratings, actually enhanced them.[24]

According to Nielsen Media Research, an average U.S. audience of only 24.5 million watched the Rio Olympics every night of the two-week event. However, about 100 million unique users streamed 3.3 billion minutes of the Games via their computers and smartphones, a higher number than ever before.[25]

Figure 8.1 illustrates the transformation taking place in television coverage of the Winter Olympics. Although it is based on data for only the Winter Olympics, the same changes took place in coverage of the summer games. The rate of increase in hours of linear television coverage slowed just as hours of digital coverage rapidly increased. In Pyeongchang broadcasters covering the Olympics produced 60,771 hours of television coverage, which amounted to a 12 percent increase over Sochi, and 97,041 hours of digital content, which was a 62 percent increase over Sochi.

Social media change the Olympic experience

Both London in 2012 and Sochi in 2014 have been characterized in the media in terms of the revolutionary changes brought about by the arrival of mobile broadband, smartphones and social media. In Sochi, the IOC put up its own social media platform called "The Olympic Athletes' Hub," which combined feeds from more

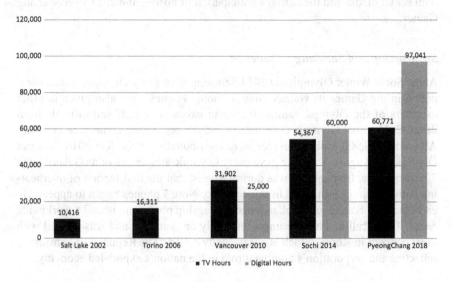

Figure 8.1 Evolution of Winter Olympics coverage hours

Source: IOC reports.

Table 8.1 Top teams ranked by social media posts (Facebook, Twitter and Instagram) in Sochi 2014

Country	Posts	Country	Posts
1 U.S.A.	22,598	11 Italy	2,060
2 Canada	15,716	12 Norway	1,622
3 Great Britain	9,867	13 Netherlands	1,553
4 Australia	5,233	14 New Zealand	1,468
5 France	4,950	15 Austria	1,256
6 Finland	4,030	16 Sweden	1,250
7 Germany	3,252	17 South Korea	869
8 Spain	2,180	18 Czech Republic	808
9 Slovenia	2,115	19 Slovakia	779
10 Switzerland	2,109	20 Uzbekistan	675

Source: International Olympic Committee. (n.d.). *Sochi 2014 social media metrics: Hot numbers. cool conversation. all yours.*

than 6,000 Olympians on Instagram, Facebook and Twitter.[26] Table 8.1 shows the top national teams at Sochi ranked by their social media use.

Two patterns stand out in Table 8.1. First, 18 of the top 20 teams are Western countries with a longer winter sports tradition. Second, South Korea was the only East Asian nation in the top 20, probably reflecting both its athletes' familiarity with social media and the nation's anticipation of hosting the 2018 Pyeongchang Games.

Sponsorship and Samsung's challenge

At the Sochi Winter Olympics in 2014 Samsung gave every Olympic athlete competing in the Games its Galaxy Note 3 phone. Phones were also given to other members of the Olympic family, including executives, staff and officials from the IOC, National Olympic Committees and the organizing committee in Sochi. Altogether 18,000 mobile devices were distributed.[27] At the Rio 2016 Summer Olympics in August, Samsung gave every Olympic athlete one of its Galaxy Edge 7 smartphones. Ironically, it was during August that the first reports of overheating batteries, smoke and fires in some Galaxy Note 7 phones began to appear. By early October Samsung had abandoned its flagship phone and issued a worldwide recall. The recall had a huge impact not only on Samsung and was viewed with great concern in Korea, which is occasionally called "The Republic of Samsung," reflecting the corporation's dominant role in the nation's export-led economy.

The new network architectures

Although the Olympic Games in Seoul required a substantial management network infrastructure, the networking requirements for the modern Olympics took a major leap forward with the arrival of smartphones and social media and the global trend toward all IP networks. Olympic networks now regularly pose a challenge for the vendor selected to design and build them and foreshadow the shape of future networks around the world. The challenge revolves around the simple fact that everyone attending the Olympics these days is carrying one or more mobile devices. The demands on wireless networks for the "bring your own device" Olympics are dramatically different than the older fixed networks. The network architect for Avaya noted that wireless communication was not a big factor in the 2010 Vancouver Games, with wired access points exceeding wireless by about a 4:1 ratio. He expected that ratio to reverse in Sochi 2014.[28]

Avaya, the vendor for the 2006 Winter Olympics in Torino and the 2010 Winter Olympics in Vancouver built the first all-IP network to support the Olympics. Avaya was also the vendor for the Sochi Olympics in 2014, which marked the first time network fabrics, virtual networks and IPTV were deployed in the Olympics.[29] The nuts and bolts structure of the Sochi network included the following:

- 54 Terabit backbone
- 2,000 Ethernet switches
- 50,000 Ethernet ports
- 2,500 wireless access points
- 36 HD channels
- 1,500 IPTV screens
- 6,500 VOIP phones

The network was constructed to support eleven competition venues, two data centers, two operation centers, three Olympic villages, two media centers and two celebration centers. It served approximately 50,000 network users, comprised of 5,500 athletes, 14,000 media and 25,000 volunteers, plus fans in attendance at the games.[30]

5G in Pyeongchang

Intel, Samsung and KT partnered to provide 5G services at the Pyeongchang Winter Olympics. The telecommunications network for the Games, built by KT, included a transport network, LTE Network and broadcasting network. The communication duct lines for the network stretched to nearly 1,400 kilometers in length and contained over 35,000 wired lines. The network included over 5,000 Wi-Fi access points to support both 4G and 5G access and the mobile communication network was capable of supporting up to 250,000 concurrently active devices.[31]

The IOC promoted the Pyeongchang 2018 experience as the "world's first broad scale 5G network" and Korean companies, led by KT, lost no opportunity to tout their 5G capabilities. The IOC claimed that "Spectators will enjoy the Olympic winter games more vividly than ever before as 5G is a game-changer for video, allowing viewers to enjoy high resolution media streaming at gigabit speeds and low latency."[32] Actual 5G services during the Pyeongchang games included the following:

- One hundred cameras placed around the Olympic Ice Arena captured 360-degree real-time views of the action on the rink and transmitted it to nearby edge servers. That information was then sent to KT's Olympic 5G network data center, where servers using Intel scalable processors rapidly produced time-sliced views of the athletes in motion. It was then transmitted over Intel's 5G Mobile Trial Platform providing gigabit speed connectivity.
- A secondary time-slicing demonstration was available in the KT "5G Connected." Pavilion, which allowed attendees to step onto a virtual version of the Gangneung Ice Arena and show off a few of their ice hockey moves.
- Multiple 5G-connected cameras were set up along the cross-country course in Alpensia, capturing the skiers as they traveled along their path. On the slopes, omni-view and multi-angle broadcasting technology is placed to provide personalized views of athletes in motion, triggered by GPS sensors.
- The "Gwanghwamun KT live site" provided an Olympic atmosphere for people in Seoul unable to attend the games in Pyeongchang, offering a 5G device experience zone including various 5G programs such as "Bobsleigh Challenge" – a motion sensor game to experience bobsled racing.[33]

The 5G experiences also included a VR-enhanced ski jumping and snowboarding experiences, which KT promoted at trade fairs and exhibitions in the years leading up to the Pyeongchang Olympics. There were also events featuring live holograms powered by 5G-based, real-time streaming technology.

However, the actual event of the 2018 Olympic Games fell short of a full implementation of 5G for two major reasons. First, the athletes, spectators and members of the Olympic family did not yet have 5G compatible mobile devices, except as provided in special experience zones as described above. This reality alone placed a clear constraint on the degree to which 5G was actually implemented. Nevertheless, the Pyeongchang games did provide multiple opportunities for people to experience and glimpse the future possibilities presented by 5G.

A second major qualification that needs to be added when describing Pyeongchang as the "5G Olympics" has to do with 5G standards. There could be no guarantee that all of the solutions introduced by Korea would be accepted as global standards for 5G, which were scheduled for final approval by international standards groups in 2020, three years following Pyeongchang.

The challenge to meet global standards

Over the past quarter century, the International Telecommunications Union developed the IMT (International Mobile Telecommunication system) framework of standards for mobile telephony and it continues to lead international efforts involving governments and industry players to produce the next generation standards for global mobile communications. Interestingly, although the ITU itself plays a central role in setting global standards, it acknowledges directly and indirectly the problems with using a versioning approach to describe technology as 3G, 4G, 5G and so forth. Its online FAQ, last updated in 2018, stated that "ITU does not have a definition for "4G" and ITU cannot hold a position on whether or not a given technology is labelled with that term for marketing purposes."[34]

Simply put, the 5G network technology demonstrated to the world in Pyeongchang 2018 could be modified by international standards bodies before the final global standards are issued in 2020. This made for a somewhat tricky and risky situation for KT. To deal with this risk, KT founded a group called "PyeongChang 5G SIG" that included major global vendors, to work on Pyeongchang 5G specifications.

A typical standardization process for international mobile telecommunications at the ITU starts with discussion of visions, followed by recommending visions that include key objectives and parameters. Despite its intent to select just a few, easy to understand and intuitive parameters, the initial discussions of 5G standardization included more parameters than considered earlier in 4G standardization.

> For instance, only two parameters – mobility and peak data rate – were included in 4G vision. In 5G vision, however, three usage scenarios – Enhanced Mobile Broadband, Massive machine type communications, and Ultra-Reliable and Low Latency communications – were proposed. As a result, other parameters that satisfy the additional scenarios were added in 5G vision.[35]

Olympic diplomacy

Just as they had in Seoul three decades earlier, the Pyeongchang 2018 Winter Olympics played an important political and diplomatic role. On January 9, 2018, North Korea agreed to attend the games in Pyeongchang. The last-minute agreement was reached in talks with South Korea after the North Korean leader Kim Jong Un had mentioned the possibility of Olympic participation in his new year's speech.

North Korea sent a small group of athletes along with a much larger delegation that included cheerleaders and high-level government officials. At the Pyeongchang opening ceremony, North and South Korean athletes marched in together under the Korean unification flag. The high-level delegation from the North included Kim Jong Un's sister, Kim Yo Jong and Kim Yong Nam, a high-level government official.[36]

The talks between North and South Korea on the occasion of the 2018 Winter Olympics led to three Inter-Korean summit meetings later in 2018 and one in Singapore between the United States and North Korea. Those meetings, especially the North-South ones, led to some tangible results. However, as of this writing the ultimate outcome of the inter-Korean diplomacy initiated by the Pyeongchang Olympics is still uncertain, and there are many contentious issues. However, what is certain is that once again, three decades after the politically momentous 1988 Seoul Olympics, the Winter Olympics in Pyeongchang played a powerful role in Korean politics. That role has implications not only for the two Koreas and eventual unification, but for the stability and future of the Northeast Asian region and its role in the world.

The legacy of Pyeongchang

The Pyeongchang Winter Olympics set a new benchmark in the transformation of the Olympics from a global mass media event to a participatory, network era, social media occasion. Athletes, coaches, members of the Olympic family and spectators in Korea experienced new perspectives made possible by the high speeds and low latency of 5G. A great deal of anecdotal evidence from industry publications indicated the manner in which the new digital networks built for the Pyeongchang games influenced the development of 5G networks and services. Pyeongchang established some precedents for media and digital coverage that would be built upon subsequent Olympic Games in the Asia region and around the world. Finally, the many Korean companies that built and operated the 5G network infrastructure for Pyeongchang benefitted in terms of both international exposure and collaboration in shaping the final global standards for 5G. It seemed likely that Korea could achieve its goal of being the first country in the world to have an operating nationwide 5G network.

Notes

1 Mobile World Live. (2016). *How will the Olympics shape 5G?* Mobile World Live, p. 2. Retrieved December 25, 2016, from www.mobileworldlive.com/whitepaper-how-will-the-olympics-shape-5g/

2 Mobile World Live. (2016). *How will the Olympics shape 5G?* Mobile World Live, p. 5. Retrieved December 25, 2016, from www.mobileworldlive.com/whitepaper-how-will-the-olympics-shape-5g/

3 Mobile World Live. (2016). *How will the Olympics shape 5G?* Mobile World Live, p. 7. Retrieved December 25, 2016, from www.mobileworldlive.com/whitepaper-how-will-the-olympics-shape-5g/

4 Dayan, D. A. (1992). *Media events: The live broadcasting of history.* Cambridge, MA: Harvard University Press, p. 1.

5 de Moragas Spa, M. N. (1995). *Television in the Olympics.* London: John Libbey, pp. 209–221.

6 Larson, J. F. and H. S. Park. (1993). *Global television and the politics of the Seoul Olympics.* Boulder, CO: Westview Press, p. 12.

7 Larson, J. F. and H. S. Park. (1993). *Global television and the politics of the Seoul Olympics.* Boulder, CO: Westview Press, p. 2.

8 Larson, J. F. and H. S. Park. (1993). *Global television and the politics of the Seoul Olympics.* Boulder, CO: Westview Press, p. 12.

9 Larson, J. F. and H. S. Park. (1993). *Global television and the politics of the Seoul Olympics.* Boulder, CO: Westview Press, pp. 171–187.

10 Larson, J. F. and H. S. Park. (1993). *Global television and the politics of the Seoul Olympics.* Boulder, CO: Westview Press, p. 95.

11 Larson, J. F. and H. S. Park. (1993). *Global television and the politics of the Seoul Olympics.* Boulder, CO: Westview Press, p. 99.

12 Larson, J. F. and H. S. Park. (1993). *Global television and the politics of the Seoul Olympics.* Boulder, CO: Westview Press, pp. 103–104.

13 Larson, J. F. and H. S. Park. (1993). *Global television and the politics of the Seoul Olympics.* Boulder, CO: Westview Press, p. 104.

14 Butler, N. (2014, August 17). *Samsung becomes 11th company to extend TOP sponsorship of IOC through to 2020.* Retrieved December 28, 2016, from Inside the Games: www.insidethegames.biz/articles/1021941/samsung-becomes-latest-company-to-extend-top-sponsorship-of-ioc-through-to-2020

15 Retrieved November 13, 2018, from www.olympic.org/ioc-financing-revenue-sources-distribution

16 This section draws heavily on Chapter 5, Korea enters the information age: The Olympics and telecommunications. In James F. Larson and Heung-Soo Park, *Global television and the politics of the Seoul Olympics.* Boulder, CO: Westview Press, 1993.

17 (1988, September). With the blink of an eye: Will Korea hitch its future to the telecommunications revolution? *Korea Business World*, 4, p. 25.

18 Mobile World Live. (2016). *How will the Olympics shape 5G?* Mobile World Live. Retrieved December 25, 2016, from www.mobileworldlive.com/whitepaper-how-will-the-olympics-shape-5g/

19 PyeongChang 2018 global broadcast and audience report, June 2018, pp. 4–6. Retrieved November 13, 2018, from https://stillmed.olympic.org/media/Document%20Library/OlympicOrg/Games/Winter-Games/Games-PyeongChang-2018-Winter-Olympic-Games/IOC-Marketing/Olympic-Winter-Games-PyeongChang-2018-Broadcast-Report.pdf

20 Riseley, M. (2012). *The multi-screen Olympics.* London: Thinkinsights Google, p. 6. Retrieved December 27, 2016, from www.thinkwithgoogle.com/intl/en-gb/research-studies/multi-screen-olympics.html

21 Nielsen. (2008, September 5). Press release. *The final tally – 4.7 billion tunes in to Beijing 2008–more than two in three people worldwide: Nielsen.* Retrieved July 25, 2016, from www.nielsen.com/content/dam/corporate/us/en/newswire/uploads/2008/09/press_release3.pdf

22 International Olympic Committee. (n.d.). *Broadcasters.* Retrieved July 28, 2016, from www.olympic.org: www.olympic.org/broadcasters

23 Bassam, Tom. (2018, February 19). All eyes on PyeongChang: How the Olympic broadcasters are changing the ratings game. *SportsPro.* Retrieved November 13, 2018, from www.sportspromedia.com/analysis/pyeongchang2018-olympic-broadcast-ratings-total-video

24 International Olympic Committee. (2009, September). *Audience report Beijing: Games of the XXIX Olympiad, Beijing 2008 global television and online media report*, p. 3. Retrieved July 28, 2016, from www.olympic.org/broadcasters: https://stillmed. olympic.org/media/Document%20Library/OlympicOrg/Games/Summer-Games/ Games-Beijing-2008-Olympic-Games/IOC-Marketing-and-Broadcasting-Various-files/Global-Television-and-Online-Media-Report-Beijing-2008.pdf#_ga=1.2264262 02.1118250054.1469508

25 Lieneck, A. (2016, August 27). NBC's digital ratings show the future of Olympics coverage. *Sports Illustrated*. Retrieved December 27, 2016, from www.si.com/tech-media/ nbc-rio-olympics-games-tv-ratings-digital#

26 International Olympic Committee. (n.d.). *Sochi 2014 social media metrics: Hot numbers. cool conversation. All yours*. Retrieved July 28, 2016, from www.olympic.org/ news: www.olympic.org/news/sochi-2014-social-media-metrics-hot-numbers-cool-conversations-all-yours

27 *Sochi sponsor profile TOP Sponsor – Samsung*. (n.d.). Retrieved December 28, 2016, from www.sponsorship.com: www.sponsorship.com/Latest-Thinking/Sochi-2014-Olympic-Sponsorship-Insights/Samsung.aspx

28 Leaks, S. (2014, February 7). *How Avaya will keep Sochi connected for the 2014 Winter Olympics*. Retrieved from Sport Techie: www.sporttechie.com/2014/02/07/ how-avaya-will-keep-sochi-connected-for-the-2014-winter-olympics/

29 Snyder, B. (2014, February 10). *Inside the network powering the Sochi Olympics: The Winter Olympics in Sochi, Russia, requires an innovative network*. Retrieved July 29, 2016, from Network World: www.networkworld.com/article/2226324/cisco-subnet/ inside-the-network-powering-the-sochi-olympics.html

30 Snyder, B. (2014, February 10). *Inside the network powering the Sochi Olympics: The Winter Olympics in Sochi, Russia, requires an innovative network*. Retrieved July 29, 2016, from Network World: www.networkworld.com/article/2226324/cisco-subnet/ inside-the-network-powering-the-sochi-olympics.html

31 Netmanias. (2016, March 13). *How ready is KT for 5G Olympics?* Retrieved July 29, 2016, from Netmanias: www.netmanias.com/en/?m=view&id=blog&no=8499

32 IOC. (2018, February 9). *Fans of the Olympic winter games 2018 to experience world's first broad-scale 5G network*. Retrieved November 15, 2018, from www.olympic.org/ news/fans-of-the-olympic-winter-games-2018-to-experience-world-s-first-broad-scale-5g-network

33 IOC. (2018, February 9). *Fans of the Olympic winter games 2018 to experience world's first broad-scale 5G network*. Retrieved on November 15, 2018, from www.olympic. org/news/fans-of-the-olympic-winter-games-2018-to-experience-world-s-first-broad-scale-5g-network

34 ITU. (2013, May 29). *International telecommunication union-radiocommunication sector ITU-R FAQ on International Mobile Telecommunications (IMT)*, p. 4. Retrieved July 30, 2016, from www.itu.int: www.itu.int/en/ITU-R/Documents/ITU-R-FAQ-IMT.pdf

35 You, H. R. (2015, March 6). *Key parameters for 5G mobile communications (ITU-R WE 5D Standardization Status*, p. 1. Retrieved July 30, 2016, from Netmanias: www. netmanias.com/en/?m=view&id=blog&no=7335&tag=589&page=2

36 Wikipedia. *North Korea at the 2018 Winter Olympics*. Retrieved November 15, 2018, from https://en.wikipedia.org/wiki/North_Korea_at_the_2018_Winter_Olympics

9 Next generation networks

The convergence of 5G and public safety LTE networks

The 2018 Pyeongchang Winter Olympics provided ample reason for Korea to display its next generation (5G) digital network technologies to the world. However, Korea had another cogent reason: its commitment to build a dedicated nationwide public safety (PS-LTE) network. It would very likely be the world's first nationwide operating public safety network, preceding similar efforts by the U.S., Canada, the UK and other nations, just as its commercial 5G mobile networks would most likely be the world's first to operate on a nationwide basis.

Korea's commitment to both the 5G and PS-LTE networks represented the latest wave of innovation in its continual efforts to upgrade and advance the nation's information superhighways. As such these projects provide a fascinating glimpse into the future of digital networks and into Korea's role as a "testbed for the world." They offer an ideal theme for this penultimate chapter as the policies, planning and construction of commercial 5G networks around the world converge at many points with the plans of nations and leading corporations to build LTE-based public safety networks.

Changing perspectives on disasters and communication

As Harold Lasswell noted in his classic article on the structure and function of communication in society, communication serves two important functions in a society:

- surveillance of the environment and
- correlation of the parts of society in responding to the environment.[1]

Although the digital network revolution has transformed societies around the globe, these two functions persist. What has changed is that digital networks offer an array of new modes of communication based on exponential increases in the human ability to store, compute and communicate information via small, powerful mobile devices. While the social functions may not have changed a great deal, the technical capabilities and possibilities presented by digital networks are nothing short of revolutionary.

Consider the role of communication in relation to current thinking about the disaster management cycle. Variations on the model presented are now widely circulated. The basic model is circular including the following four stages in clockwise sequence after a disaster strikes:

- Response
- Recovery
- Mitigation
- Preparedness

One has only to think back to the origins of the internet itself, as a network designed in part by the U.S. military that would be able to survive attack. One study suggested that the design of both ARPANET and the internet "favored military values, such as survivability, flexibility and high performance over commercial goals such as low cost, simplicity or consumer appeal."[2] Indeed, a growing accumulation of experience with internet-based mobile communications underscores the important function of mobile communication in a variety of disaster scenarios, both natural and human-caused.

As noted in a 2015 UN report on the future of disaster risk management,

> Interpreting disasters as exogenous shocks lies at the root of the disaster management cycle, which – as its name implies – revolves around disasters as events. While the disaster management cycle was and still is seductive due to its simplicity and internal logic, it encouraged and justified the syncretic expansion of emergency management organizations into other aspects of disaster risk management, such as prevention, reduction and recovery.[3]

Ultimately, the report argues

> this approach to disaster risk reduction encapsulates a fundamental contradiction: it aims to protect the same development paradigm that generates risk in the first place. As such, if increased investments are made to protect development without addressing the underlying risk drivers at the same time, more and more effort will lead to diminishing returns and flagging progress. Disaster risk will continue to be generated faster than it can be reduced."[4]

More ominously the report warns that

> given the growing evidence of systemic risk at the planetary scale, there is now a very real possibility that disaster risk will reach a tipping point beyond which the effort and resources necessary to reduce it will exceed the capacity of future generations.[5]

The changing character of disasters and risk

During the 20th century mass media era sociologists and communication scholars recognized the important role played by the media in relation to both natural and human-caused disasters. Indeed, a committee was formed by the National Academy of Sciences in 1979 to examine that role.[6] However, since that time globalization and the rise of digital networks led by the internet, contributed to a broadened conception of disasters.

In the hyperconnected digital network era, two of the secular trends emphasized in this book shaped a new and broader understanding of disasters. One was the global phenomenon of terrorism in which the events of September 11, 2001, marked a dividing line in history and the start of a new kind of war, which has been waged in the cities of Europe as well as in Syria and Iraq. The other long-term trend was the growing public awareness and the global political commitment to combatting climate change. In today's world, terrorist attacks and all the consequences of the war against terrorism, including massive migrations of refugees can all be considered types of disasters. Likewise, the long-term trends such as increased carbon emissions, depletion of the ozone layer surrounding the earth and other measures of unsustainable practices contributing to climate change constitute a new form of disaster on a truly global scale.

The Hyogo Framework for Action 2005–2015, adopted by the World Conference on Disaster Reduction held in Kobe, Hyogo, Japan, in January 2005 made the relationship between disaster risk reduction and sustainable development quite explicit. It identified the following five priorities for action, each of which implies important roles for information, knowledge and the new digital networks:

- Ensure that disaster risk reduction (DRR) is a national and a local priority with a strong institutional basis for implementation.
- Identify, assess and monitor disaster risks and enhance early warning.
- Use knowledge, innovation and education to build a culture of safety and resilience at all levels.
- Reduce the underlying risk factors.
- Strengthen disaster preparedness for effective response at all levels.[7]

The subtitle of the 2015 United Nations global assessment report on disaster risk reduction, *Making Development Sustainable: The Future of Disaster Risk Management*, says it all. That report notes that

> Growing disaster risks, climate change as well as poverty and inequality are all indicators of unsustainability. . . . In this context, disaster risk reduction is at a crossroads: It can continue to focus on managing an increasing number of disasters or it can shift the focus to managing the underlying risks in a way that facilitates sustainable development.[8]

The report emphasized that "Disasters have been interpreted as threatening development from the outside. As a result, disaster risk generation within development has not been addressed effectively."[9]

The UN report also traces the origins of what is now known as a "disaster risk management sector" in most countries, consisting of "the institutions, legislation and policies, administrative arrangements and instrumental systems created to respond to and manage disasters and crises." Although such sectors have a longer history, in the United States the Federal Emergency Management Agency (FEMA) was assimilated into the Department of Homeland Security following the September 11, 2001, attacks in New York and Washington, DC.[10]

The UN report describes a slow evolution at both national and international levels, from managing disasters to managing risks. It notes that disaster risk management began to emerge in the 1970s as a specialized domain and sector.[11]

Disasters as an impetus for PS-LTE network development

Following the September 11, 2001, terrorist attacks a bipartisan National Commission on Terrorist Attacks Upon the United States (also known as the 9–11 Commission) was formed to prepare a complete account of the circumstances surrounding the attacks, including preparedness, immediate response and recommendations to guard against future attacks. The Commission's public report, released on July 22, 2004, contained the following comments:

> The inability to communicate was a critical element at the World Trade Center, Pentagon, and Somerset County, Pennsylvania, crash sites, where multiple agencies and multiple jurisdictions responded. The occurrence of this problem at three very different sites is strong evidence that compatible and adequate communications among public safety organizations at the local, state, and federal levels remains an important problem.[12]

On July 7, 2002, a report in *The New York Times* described the situation more graphically in the following manner:

> Minutes after the south tower collapsed at the World Trade Center, police helicopters hovered near the remaining tower to check its condition. "About 15 floors down from the top, it looks like it's glowing red," the pilot of one helicopter, Aviation 14, radioed at 10:07 a.m. "It's inevitable." Seconds later, another pilot reported: "I don't think this has too much longer to go. I would evacuate all people within the area of that second building." Those clear warnings, captured on police radio tapes, were transmitted 21 minutes before the building fell, and officials say they were relayed to police officers, most of whom managed to escape. Yet most firefighters never heard those warnings, or earlier orders to get out. Their radio system failed frequently that morning. Even if the radio network had been reliable, it was not linked to the police system. And the police and fire commanders guiding the rescue efforts did not talk to one another during the crisis.[13]

The recommendation of the 9–11 Commission was that Congress should support pending legislation providing for the expedited and increased assignment of radio spectrum for public safety purposes. It also suggested that high-risk urban areas should establish signal corps units to ensure communications connectivity between among civilian authorities, local first responders and the National Guard and that federal funding of such units should be given high priority.[14]

In 2012 the U.S. Congress finally acted, creating The First Responder Network Authority or FirstNet for short. Its website makes reference to the 9–11 Commission, noting that the "The final recommendation of the 9–11 Commission, the network will connect police officers, firefighters and EMS providers like never before."[15] Thus, from its inception, FirstNet was centrally concerned with the needs of police, firefighters and other first responders.

Just as a disastrous terrorist attack moved the U.S. government toward implementation of a nationwide dedicated public safety network, disasters in South Korea provided an impetus for the nation to act. In April of 2014 a ferry traveling from Incheon to Jeju Island and carrying hundreds of high school students on a field trip capsized and sank, killing most of the students.

It would be no exaggeration to say that this tragic incident galvanized public support for action and in one fell swoop eliminated political obstacles at many levels. In the immediate aftermath of the Sewol tragedy, the need for a unified public safety network was brought up in the national assembly, and the government initiated a project to construct such a network a month later. In July of 2014 the Ministry of Science, ICT and Future Planning selected Public Safety LTE (PS-LTE) as the technology for the new network, and in November of the same year the Ministry of Public Safety and Security started implementation with the aim of constructing the nationwide network by the end of 2017.[16] That initial aim turned out to be overly ambitious as the costs and complexities of the project mounted.

The SafeNet Forum was also created in 2014 as an official body to reflect the opinions of various parties and to monitor the government's policies in order to maintain fairness and transparency in the process of completing construction of a nationwide public safety network.[17] The forum included all of the major players from industry, academic and government. The four ministries involved, as of early 2016 were the Ministry of Public Safety and Security, the Ministry of Science, ICT and Future Planning, the Ministry of Oceans and Fisheries and the Ministry of Land, Infrastructure and Transport. The latter two ministries were added to the SafeNet Forum in early 2016, reflecting the goal of integrating the first-stage PS-LTE network with an LTE-R network for high-speed rail and an LTE-M network for maritime traffic in the ocean surrounding South Korea's coastline.

Korea's goals for next generation networks

At the ITU Plenipotentiary conference in Busan in November of 2014 Korea's President Park Geun Hye publicly stated that

> We . . . have reached an inflection point in the hyper-connected digital revolution – a revolution defined by "increased connection, smarter connection, and faster connection." This hyper-connected digital revolution will lead to new converged industries and services such as smart vehicles, smart

healthcare, and smart cities. It will transform the way we live and contribute to the development of our economies and societies.

She continued with some very specific comments about Korea's near-term goals for broadband networks, pledging that

To lead the hyper-connected digital revolution, Korea plans to build a nation-wide Giga Internet network by 2017. We also are focusing on technological development and infrastructure upgrade with the aim of being the first country to launch commercial 5th generation mobile communication services in 2020.[18]

South Korea is well on its way to achieving both of these goals, thanks in part to two very different events, the awarding of the 2018 Winter Olympic Games to the city of Pyeongchang and the tragic sinking of the Sewol Ferry after which the nation accelerated its plans for construction of a nationwide PS-LTE network.

Next generation networks, widely referred to as 5G, will move storage, computing and communications capacity closer to the edge of the network. One of the earliest and most ambitious effort to strengthen capabilities at the edge of the network was Cisco's fog computing platform. It represented a fundamental shift in internet architecture that includes hardware, including a new generation of "edge routers" that link access networks to the core of the internet, and software.

The fog metaphor, referring to a cloud close to the ground, is highly appropriate in this era of rapidly proliferating mobile devices, remote sensing and wearable devices leading toward the Internet of Everything (IoE). Fog devices are denser and closer geographically to users. They are also more heterogeneous than cloud devices, ranging from end user devices, to access points to routers and switches. The fog does not replace cloud computing, but rather enables a new set of applications and services with an interplay between the cloud and the fog. As defined by Cisco researchers, "Fog Computing is a highly virtualized platform that provides compute, storage and networking services between end devices and traditional Cloud Computing Data Centers, typically, but not exclusively located at the edge of the network."[19]

In Korea KT's plan for 5G service assumes several related changes will take place as follows:

- If the core network moves closer to cell sites, application servers will follow.
- Backhaul traffic will significantly decrease.
- Although such services as real-time remote control or autonomous vehicles may require less traffic than video, they will require ultra-low delays (latency).

The Pyeongchang Winter Olympics were held in February 2018, before international standards on 5G Radio Access Technology (RAT) are finalized. Therefore, KT founded a group named Pyeongchang 5G SIG in collaboration with global vendors to work on "Pyeongchang 5G" specifications. Given its investment in Pyeongchang, KT obviously hoped that these specifications would be the same as those approved by the relevant international standards organizations in 2020.[20]

Korea's PS-LTE networks

The size and possible impact of Korea's PS-LTE projects are difficult to pin down but can be estimated from the level of business interest in such networks, in Korea, the U.S., Canada, the UK and elsewhere. A news report in January 2015 reported intense competition among global communications equipment suppliers for participation in the Korea project. It noted that

> According to sources in the government and communications industry on Jan. 25, the national disaster safety communications network project is esti-mated to be worth 2 trillion won (US$1.85 billion). However, the size of the project is expected to increase to more than 3 trillion won (US$2.8 billion) if 10 year-maintenance costs are included. So competition between global com-munications equipment suppliers for the project is heating up.[21]

By comparison, a report by the Congressional Research Service in the United States estimated that the cost of building a nationwide wireless network for public safety would be in the tens of billions of dollars. In Korea, since the PS-LTE pro-ject was announced, the press has carried numerous reports of alliances between Korea's manufacturers and service providers and their international counterparts.

Many of the technical characteristics required for 5G mobile networks are also needed for the new generation of public safety LTE networks. Therefore, it was not surprising at all that Pyeongchang, the host city for the 2018 Winter Olympics, and the surrounding venues in nearby cities, became the pilot phase, completed in 2015, for the new PS-LTE network. The network construction project literally gave Pyeongchang in Gangwon province and some neighboring cities some of the most advanced broadband networks in the entire world.

The basic, dedicated PS-LTE network is only the first stage, with plans for mari-time and railway networks to be integrated at a later date. As of this writing, techni-cal work and pilot testing for the maritime and railway networks is well underway.

Next generation public safety networks: commercial 5G and PS-LTE convergence

Current public safety networks are based on land mobile radio (LMR) systems such as terrestrial trunked radio (TETRA). These legacy systems allow a number of standard services that are very important for public safety communications. Consequently, when the transition to LTE public safety networks is complete, the future networks will need to provide the following services:

- Group communications
- Device-to-device communication

- Push-to-talk (PTT) or mission-critical-push-to-talk (MCPTT) service
- Prioritization and pre-emption
- Emergency calls

In addition to the above services, future public safety networks need to have the following characteristics:

- Geographic coverage close to 100 percent
- Scalability
- Availability (resilience and high service availability)
- Security
- Interoperability (with other public safety networks)[22]

Although the construction of a nationwide, dedicated public safety LTE network may at first glance appear to be a separate project from the building of a 5G network, in Korea and other countries undertaking such efforts, the two types of networks are converging. For example, the mission-critical push-to-talk (MCPTT) feature, which includes group calling/broadcast capabilities and direct-mode functionality that supports communication when the LTE network is unavailable, may have started as a public safety initiative. However, the commercial sector now sees the potential of this functionality as a means to more efficiently move traffic on capacity-strained networks and to support communication between IoT devices. There are other examples, including resilience and reliability. First responders will, of course, require hardened communications that work at all times, even during crises. However, the emerging reality is that the public's growing reliance on wireless communication for both routine and emergency calls makes the same feature desirable in commercial networks. In summary, the convergence between public safety and commercial networks is a two-way street.[23]

As a May 2018 report by Ericsson put it,

> The idea of using commercial technology for mission-critical, emergency services has not been successful in the past because commercial technology lacked a number of important capabilities. Today, however, all the necessary technology is in place, namely: high-capacity radio (LTE), an evolution path to even more capable radio (5G), virtualization, software defined networking, automation, network slicing, positioning, proximity communication, the ability to prioritize communication and secure data transfer.[24]

The growing international commitment to build public safety LTE networks grew out of a realization on the part of the public, policymakers and first responders that commercial LTE networks already possessed capabilities that were not available in the patchwork quilt of non-interoperable emergency networks. The following pages examine some of those rapidly emerging possibilities.

Real-time mobile and visual communication

Across a wide range of disaster scenarios, future victims will have real-time communication capability thanks to the digital network revolution and the "law of mobility," usually attributed to Russ McGuire of Sprint. The law simply states that the value of any product or service increases with its mobility.[25] Mobility is measured as the percentage of time that a product or service is available for use. In the aftermath of certain types of disasters, such as earthquakes or the sinking of the Sewol Ferry the first minutes or hours are frequently referred to as the "golden time," during which survivors can be rescued and given proper medical treatment. In the future, LTE devices promise to play a new role by allowing victims to communicate not only with friends or family, but with nearby victims and with first responders by voice, text or video.

Not only victims and first responders in a disaster situation, but also government officials responsible for public safety more broadly, might benefit from the capabilities of the new LTE networks. For example, the disastrous 2003 subway fire in Daegu, which was captured on surveillance video, spurred additional research funding in Korea for research on the development of facial recognition or other algorithms that could alert public safety officials in real time about any possible threats.

When the powerful Hurricane Michael struck the Southeastern United States in October 2018 FirstNet reported that its mobile network operated at 90 percent or better of normal performance in areas affected by the storm. AT&T Communications CEO John Donovan attributed this performance to preparation for the storm and use of deployable assets in the affected areas, noting that "We are seeing, in real time, how we're performing in times of emergency, with Hurricane Michael being the latest example."[26]

Future public safety and commercial networks based on LTE, and eventually 5G will not only provide real-time information. Some of the information will come in the form of video or visualizations that can save lives and be a tremendous aid to first responders. In general, people around the world prefer video to all other forms of digital information. As of late 2018 the Cisco Visual Networking Index (VNI) forecast projected that 82 percent of all IP traffic would be video by the year 2022.[27]

Public safety organizations and first responders will benefit from having digital smart maps of buildings, with all their assets, systems and sensors geolocated. Such visual information can be quickly assimilated by first responders and could be presented as a form of augmented reality (AR). Future buildings will be able to identify threats and create such smart maps, allowing the problem to be dealt with before it becomes a disaster. First responders, either enroute or at the scene would have a new level of situational awareness, able to see what's going on inside the building before entering it, to help determine the most efficient and safest response.[28]

Interactive, group communication and proximity services

The new public safety digital networks promise to open up numerous possibilities before, during and after all kinds of disasters. These include use of the live broadcast mode, which is a standard option in the latest version of Android, or simply connecting by voice, text or video with first responders or other survivors located nearby. As of 2013 the Third Generation Partnership Project (3GPP), a global standards organization, had agreed on two main areas of LTE enhancement to address public safety applications:

- Proximity services that identify mobiles in physical proximity and enable optimized communications between them.
- Group call system enablers that support the fundamental requirement for efficient and dynamic group communications operations such as one-to-many calling and dispatcher working.[29]

Proximity services include various types of machine to machine (also called device to device) communication. During disasters, telecommunications infrastructures are often congested or even destroyed. In such circumstances an ad hoc network using Bluetooth or Wi-Fi to link smart handheld devices without relying on centralized cellular networks or wireless access points could be very valuable. Such ad hoc networks might aim to (1) accurately locate survivors, (2) help rescuers to find victims quickly and (3) allow power saving on smart devices.[30]

Location awareness

Location plays an important role in emergency response situations. Awareness of where one is located relative to a hazard can benefit first responders, victims, bystanders and even the public at large. For example, first responders need to answer several questions. Where am I? What hazards exist on the scene? Where are other first responders? Where are the victims? The GPS and mobile computing capability of smartphones offers a powerful new means for rescuers to quickly answer such questions, possibly saving lives.[31]

Mission-critical push-to-talk

In March of 2016 the Third Generation Partnership Program (3GPP) "froze" Mission-critical push-to-talk (MCPTT) functionality as part of the LTE standard in Release 13 as part of the latest release. One report indicated that

> The new MCPTT standard supports direct-mode voice communications, a "discovery" feature that lets a user know of other users that are within direct-mode range, and a relay capability that allows an out-of-coverage user to connect to a fixed LTE network via a direct-mode connection with an in-network LTE device.[32]

Interest in providing MCPTT over LTE capability to first responders continued to grow in 2016 and 2017, but widespread adoption was still a future prospect because of standards issues and the need to manufacture and distribute handsets compatible with both the new standards and legacy mission-critical systems. MCPTT was the first in a series of mission-critical services and functionalities demanded by the market. In 2017 3GPP added the following additional mission-critical services and enhancements to its repertoire of standardized applications:

- Enhancements to MCPTT
- MCData
- MCVideo
- General framework, which facilitates standardizing additional MC services[33]

Network and information security

As cyber-attacks against all commercial networks rose in the first two decades of the 21st century, security became an increasingly important capability for all communications networks. However, it assumed a new and even higher level of significance in the context of mission-critical public safety communication networks and the emergence of long term evolution (LTE) technology. Industry sources predict that cyber-attacks will increasingly seek to penetrate mission-critical communication networks. In public safety communications, the mission depends heavily on the reliability and security of the communications network. Weak or compromised network security reduces reliability and ultimately availability. Without proper security measures in place no amount of physical reliability features such as site hardening or backup power can guarantee the high level of availability required for public safety networks.[34]

As public safety agencies migrate from Land Mobile Radio (LMR) capabilities to LTE, security becomes more important. This is especially relevant for an evolved packet core (EPC), which is based on IP transport. While EPC makes the network very efficient, it also requires security architecture to protect against the diverse threats that jeopardize mobile networks, in particular IP-related cyber-attacks. The legacy public safety systems were circuit switched and built on proprietary implementations. Today's public safety networks, as they migrate to LTE are inter-networked and rely upon IP-based networking that utilizes standard ports and protocols. The new networks bring with them asymmetrical threats that can exist either in data centers or can be initiated by a laptop user halfway around the world. The threat landscape has dramatically changed.[35]

Future public safety networks will represent critical infrastructure for nations, corporations, international organizations and citizens groups. While enhanced network connectivity and interoperability can contribute greatly to public safety, it also raises the vulnerability of these networks to cyber-attacks. This is a particular concern for Korea, which ranked fourth in the world in the total number of lost or stolen data records since 2013, behind the U.S., India and China. Also, Korea ranked second in the world after the U.S. in the ratio of data records stolen to population.[36]

Network resilience

As noted earlier, the original design of the internet was influenced in part by the military objective of having a network that could withstand a nuclear attack. Some of the features suggested for public safety LTE networks include hardened base stations and mobile truck-mounted, backpack or drone-mounted stations that can be activated in a disaster location if the regular base station is knocked out.

Such physical methods to ensure network resilience can be very costly. For example, it is probably unrealistic to harden all base stations in a network. Consequently, software-based methods of network traffic management are also being explored to increase network resilience. In public safety networks, coverage, not capacity, is paramount, especially during an outage when sites are down. Through traffic control and preemption, the service level of low-priority users can be reduced or denied, freeing up resources to restore coverage to high-priority users such as first responders in an emergency.[37]

Education and training for next generation networks

The new commercial 5G and public safety LTE networks under construction in Korea and other countries represent nothing less than a new, denser digital ecosystem. The completion of these next generation networks means, among other things, a massive need for training in how to most effectively use the new technologies.

Virtual reality (VR) technology holds great promise for educational and training efforts directed at first responders, citizens and others. In contrast to AR, VR creates an entirely new and often realistic digital environment, which can be useful for training simulations to teach new skills. VR is highly immersive and the incorporation of techniques from multiplayer online games, otherwise referred to as gamification, could make training more realistic, entertaining and effective.

Next generation public safety networks may also build upon existing experience with cloud-based services. For years now companies like Google and Facebook have offered cloud-based services in response to disasters and crises. Google's crisis response page emphasizes a number of online tools including public alerts, the Google Person Finder and Google Crisis Map. These services can all be viewed as forms of crowdsourcing, which draws on the power of the internet to accomplish tasks. For example, Google Crisis Map is designed to allow users to explore disaster-related geographic data without special software, to share the map with co-workers, media outlets and partners and to contribute to a crisis map or download data from it.[38]

First responders, government policymakers, corporate executives, NGOs and citizens will all need training on how to use the new networks. Such training will no doubt draw on the internet, MOOCs and also the mobile apps ecosystem.

As described in Chapter 5, Korea has experience with massive educational campaigns, having launched "The Informatization Education Plan for 25 Million

People," in 1999 on the heels of the Asian Economic Crisis. That experience and the nation's strong long-term commitment to education may very well help it adjust to the new digital network environment provided in the coming decades by next generation networks.

Notes

1 Lasswell, H. D. (1948). The structure and function of communication in society. In L. E. Bryson, *The communication of ideas.* New York: Harper and Row, pp. 37–51, 51.
2 Abbate, J. (1999). *Inventing the internet.* Cambridge, MA: The MIT Press, p. 5.
3 United Nations. (2015). *Global assessment report on disaster risk reduction 2015: Making development sustainable: The future of disaster risk management.* New York: United Nations, p. 35.
4 United Nations. (2015). *Global assessment report on disaster risk reduction 2015: Making development sustainable: The future of disaster risk management.* New York: United Nations, p. 35.
5 United Nations. (2015). *Global assessment report on disaster risk reduction 2015: Making development sustainable: The future of disaster risk management.* New York: United Nations, p. 36.
6 National Academy of Sciences. (1980). Disasters and the mass media. *Proceedings of the committee on disasters and the mass media workshop, February 1979.* Washington, DC: National Academy of Sciences, National Research Council, p. 301.
7 United Nations Office for Disaster Risk Reduction (UNISDR) *Hyogo Framework for Action 2005–2015.* Retrieved November 29, 2018, from www.unisdr.org/files/1217_ HFAbrochureEnglish.pdf
8 United Nations. (2015). *Global assessment report on disaster risk reduction 2015: Making development sustainable: The future of disaster risk management.* New York: United Nations, p. xvi.
9 United Nations. (2015). *Global assessment report on disaster risk reduction 2015: Making development sustainable: The future of disaster risk management.* New York: United Nations, p. 23.
10 United Nations. (2015). *Global assessment report on disaster risk reduction 2015: Making development sustainable: The future of disaster risk management.* New York: United Nations, p. 29.
11 United Nations. (2015). *Global assessment report on disaster risk reduction 2015: Making development sustainable: The future of disaster risk management.* New York: United Nations, p. 31.
12 The 9–11 Commission. (2004). *The 9–11 Commission report.* Government Commission Report, p. 397.
13 Dwyer, J. K. (2002, July 7). Fatal confusion: A troubled emergency response; 9/11 exposed deadly flaws in rescue plan. *The New York Times.*
14 The 9–11 Commission. (2004). *The 9–11 Commission report.* Government Commission Report, p. 397.
15 FirstNet. (n.d.). *The history of FirstNet and the public safety broadband network.* Retrieved July 19, 2016, from www.firstnet.gov: www.firstnet.gov/content/ firstnet-building-nationwide-public-safety-network
16 Hong, D. (2016, 5). *Greetings from SafeNet forum chair.* Retrieved June 4, 2016, from SafeNet Forum: http://safenetforum.or.kr/eng/main/main.php?categoryid=02&menuid=01&groupid=00#
17 SafeNet Forum. (2016, 6). *Background and objectives.* Retrieved June 4, 2016, from SafeNet Forum: http://safenetforum.or.kr/eng/main/main.php?categoryid=02&menuid=02&groupid=00
18 ITU. (2014). *Republic of Korea Park Geun-hye President 20 October 2014 ITU plenipotentiary conference opening ceremony.* Retrieved May 29, 2016, from ITU

Committed to connecting the world: www.itu.int/en/plenipotentiary/2014/statements/file/Pages/opening-ceremony-geun-hye.aspx

19 Bonomi, F. R. (2012). *Fog computing and its role in the internet of things. Sigcomm 2012*. Helsinki: ACM, p. 13.

20 Netmanias. (2016, March 13). *How ready is KT for 5G Olympics?* Retrieved June 6, 2016, from Netmanias: www.netmanias.com/en/?m=view&id=blog&no=8499

21 Park, J. H. (2015, January 26). Global companies fiercely compete for Korea's disaster safety communications network project. *Business Korea*.

22 Stojkovic, Milan. (2016, June). *Public safety networks towards mission critical mobile broadband networks*. Norwegian University of Science and Technology, Master of Telematics, p. 27. Retrieved November 27, 2018, from https://brage.bibsys.no/xmlui/bitstream/handle/11250/2406853/15854_FULLTEXT.pdf?sequence=1

23 Jackson, D. (2016, June 9). *Convergence between public-safety and commercial communications not just a one-way street*. Retrieved June 29, 2016, from Urgent Communications: http://urgentcomm.com/blog/convergence-between-public-safety-and-commercial-communications-not-just-one-way-street

24 Ericsson. Ensuring critical communication with a secure national symbiotic network. Retrieved from www.ericsson.com/en/white-papers/ensuring-critical-communication-with-a-secure-national-symbiotic-network

25 McGuire, R. (n.d.). *McGuire's Law*. Retrieved July 24, 2016, from mcguireslaw.com: http://mcguireslaw.com/

26 Jackson, D. "FirstNet adoption increases by almost 50% during last two months, AT&T says," *Urgent Communications*, October 24, 2018. Retrieved November 27, 2018, from https://urgentcomm.com/2018/10/24/firstnet-adoption-increases-by-almost-50-during-last-two-months-att-says/

27 Retrieved from www.cisco.com/c/en/us/solutions/service-provider/visual-networking-index-vni/index.html

28 Hernandez, Joe. (2017, August 1). The power of visualization technology for public safety. *Urgent Communications*. Retrieved November 27, 2018, from https://urgentcomm.com/2017/08/01/the-power-of-visualization-technology-for-public-safety/

29 Retrieved November 28, 2018, from www.3gpp.org/news-events/3gpp-news/1455-Public-Safety

30 Han, Jiyong and Junghee Han. (2018, March 13). Building a disaster rescue platform with utilizing device-to-device communication between smart devices. *International Journal of Distributed Sensor Networks*, 14(3), pp. 1–13.

31 Betts, Bradley J., Robert W. Mah, Richard Papasin, Rommel Del Mundo, Dawn M. McIntosh and Charles Jorgensen. Improving situational awareness for first responders via mobile computing. *Smart Systems Research Laboratory, NASA Ames Research Center*. Retrieved November 28, 2018, from https://ti.arc.nasa.gov/m/pub-archive/1076h/1076%20(Betts).pdf

32 Jackson, D. (2016, March 5). *3GPP approves standard for mission-critical PTT (MCPTT) over LTE in release 13*. Retrieved July 24, 2016, from Urgent communications: http://urgentcomm.com/3gpp/3gpp-approves-standard-mission-critical-ptt-mcptt-over-lte-release-13

33 Lair, Yannick and Georg Mayer. (2017, June 20). Mission critical services in 3GPP. *3GPP*. Retrieved November 29, 2018, from www.3gpp.org/NEWS-EVENTS/3GPP-NEWS/1875-MC_SERVICES

34 Harris, Greg. (2011). *Cyber security for long term evolution: Public safety networks today and tomorrow*. Harris Corporation, p. 2. Retrieved November 29, 2018, from www.harris.com/sites/default/files/documents/solutions_grouping/cyber-security-lte-public-safety-networks-wp.pdf

35 Harris, Greg. (2011). *Cyber security for long term evolution: Public safety networks today and tomorrow*. Harris Corporation, pp. 2–3. Retrieved November 29, 2018, from www.harris.com/sites/default/files/documents/solutions_grouping/cyber-security-lte-public-safety-networks-wp.pdf

36 Sobers, Rob. (2018, July 16). The world in data breaches. *Varonis*. Retrieved November 29, 2018, from www.varonis.com/blog/the-world-in-data-breaches/

37 Rouil, R., W. Gary, C. Gentile, N. Golmie and P. Schwinghammer. (2018, February). Increasing public safety broadband network resiliency through traffic control. *Digital Communications and Networks*, 4(1), pp. 48–57. Retrieved November 29, 2018, from www.sciencedirect.com/science/article/pii/S2352864817300500

38 Google. (n.d.). *Google crisis response*. Retrieved July 24, 2016, from www.google. org/crisisresponse: www.google.org/crisisresponse/about/resources.html

10 Korea's place in cyberspace
Lessons for a sustainable world

> "Lessons for Korea are lessons for the world."
> —November 4, 2014, speech by World Bank President
> Kim Jim Yong, Seoul, Korea

This book chronicled some key aspects of South Korea's ICT-driven development from 1980 to the present, in an effort to help scholars, policymakers and students around the world place the "Miracle on the Han" in clearer context. That crucial context helps to identify policy lessons, including both successes and failures that may guide Korea and other nations around the world into a sustainable and globally networked future.

The title of this chapter draws on an important speech delivered in Seoul on November 4, 2014, on the important topic of "Human capital in the 21st century." Kim Jim Yong's core message was that creativity and entrepreneurial energy would be essential for Korea to continue its economic growth and for global development more generally. While acknowledging Korea's successful economic development, he pointed to its shortcomings in a sweeping assessment of Korean education.

This concluding chapter summarizes some of the lessons for Korea and for the world that can be gleaned from the Korean experience. The summary begins with the emergence of cyberspace in Korea and globally and moves on to more specific issues.

Conceptions of cyberspace

In historical perspective, the global emergence of a realm called cyberspace roughly paralleled Korea's digital development. John Perry Barlow's 1996 declaration of the independence of cyberspace began as follows:

> Governments of the Industrial World, you weary giants of flesh and steel, I come from Cyberspace, the new home of Mind. On behalf of the future, I ask you of the past to leave us alone. You are not welcome among us. You have no sovereignty where we gather.[1]

Among the earliest Koreans to venture into cyberspace were those who frequented the nation's ubiquitous PC Bangs to play such online games as StarCraft, World of Warcraft or Lineage. Next came such web services as Cyworld. On the whole, Koreans began to experience cyberspace years earlier and in far greater numbers than netizens from other nations.

The word "cyberspace," which comes from combining "cybernetics" and "space," was coined by science fiction novelist and cyberpunk author William Gibson in his 1987 novel *Neuromancer*. He defined cyberspace as a "consensual hallucination." Metaphorically, the term is now widely used to describe a social setting that exists purely within a space of representation and communication. It exists entirely within an electronic, computer space, distributed across increasingly complex and fluid networks. As used in academic circles and in the activist community, cyberspace has become a de facto synonym for the internet.

Defining cyberspace is very much a matter of context. However, a common factor in virtually all definitions is the sense of place that they convey. Cyberspace is most definitely a place where you chat, explore, research and play.[2]

Cyberspace and Korea's place in it will be a vital part of the future global information society, for at least three important reasons. First, as we have documented, South Korea built the information superhighways through which one enters cyberspace almost half a decade before the United States and other advanced economies. Once the digital networks were in place two new places appeared in the expanding cyberspace. One was the space occupied by massive multiplayer online games (MMOG) beginning with StarCraft. The second expanding space was social networking, as epitomized by Cyworld.

A second reason that cyberspace has profound importance for Korea has to do with the balance between manufacturing and service industries. To date, South Korea's progress in digital development has been based largely on the manufacturing and export of hardware, including semiconductors, flat panel displays and television sets, mobile devices and network hardware. The networks and communications hardware are a necessary but not a sufficient condition for cyberspace. Three quarters or more of the global communications market has to do with software, content and service applications. These are the stuff of which cyberspace is made, and it will be vital for Korea to move seriously into these areas, with the same sort of success it has achieved to date in hardware manufacturing and export.

A third aspect of cyberspace also underscores its significance for Korea. Its inherently global scope means that it poses a set of governance challenges for governments, corporations and citizens groups. Such issues are of particular importance to countries, like Korea, that export ICT products and services to the whole world. Simply put, Korea already has too much of a stake in cyberspace to sit idly by. Rather, it must be an active participant, as befits its recent experience.

Cyberspace can be thought of as the frontier built upon the infrastructure we call broadband internet. How then do Koreans, as some of the earliest settlers in that frontier area, view cyberspace, and how does their vision compare with that of other nations or cultures? Although this subject can be a slippery one, it is important for leaders everywhere to struggle with the concept and to clarify their own vision of the future, knowledge-based society.

Mainstream media reports on cyberspace feature noisy commercial businesses, stimulating or pornographic sex and horrible crimes. By contrast, in healthy people's cyberspace there are pleasant meetings, recreation, information and knowledge to satisfy a person's mental needs. In 20th century fashion cyberspace offered a form of escape. In the 21st century if the younger generations are exhausted, mentally running dry and craving for knowledge, isn't this the place to fulfill those needs?

Young people can benefit from using cyberspace as a place of refuge. As long as netizens behave themselves mentally when they enter that space, they can find a pleasant path. It is possible to visit new places, meet new people and even find love. Together the citizens of cyberspace can become aware of knowledge they didn't have and satisfy their curiosity about things they wanted to know. All requirements for knowledge and recreation can be met satisfactorily in cyberspace.

Cyberspace has created something even more realistic than the real world. To that extent, the door to this place is wide; the climate is balmy and warm. However, today there are many obstinate people in Korea who play games day and night even to the point of dying from the exertion. In cyberspace there is ample opportunity for recreation and acquiring knowledge, yet these things cannot be more important than real life. Simply turning off the computer will bring the reality of life back and expand one's sense of control.

Korea's place in cyberspace

As noted at the outset, the invention of the transistor and Shannon's mathematical theory of communication, together with the development of digital switching, unleashed exponential increases in the human ability to store compute and communicate digital information. These developments lie at the core of the digital network revolution globally and in Korea. Although Korea had missed the opportunity provided by the industrial revolution, it was among the first nations in the world to harness the power of digital networks, completing its first nationwide PSTN in 1987.

Perhaps the most significant aspect of the digital network revolution, for Korea, is its global scope. Many of the problems now facing humanity are not limited to one nation, but rather extend around the globe. They include energy, global warming, education, healthcare and a host of other issues. Virtual reality may provide a significant means for human beings to ameliorate some of their most urgent problems. To date, it has been difficult to create a high-fidelity virtual reality experience, largely due to technical limitations on processing power, image resolution and communication bandwidth. Given the pace of technology change, those limitations should eventually disappear.

The magnitude of the changes in South Korea since 1980 is now a well-documented matter of historical record. However, those changes included the following characteristics that pose major challenges for Korea's future development:

- It was based largely on manufacturing in the ICT sector, rather than services.
- The emphasis on ICT hardware manufacturing related to a relative weakness in software. One illustration was the continued widespread use of Microsoft ActiveX software for online financial transactions long after it had been discouraged as a security threat by Microsoft itself.
- It was strongly export-led.
- Korea's large industrial conglomerates played a leading role, versus small and medium-sized enterprises.
- Korea's domestic market lagged much of the world by more than two years in the widespread adoption of mobile broadband in the form of the iPhone.

Korea was one of a small handful of nations around the world that did not quickly adopt Google for web search, relying instead on Naver, a far smaller Korean language-only web portal. This showed the power of language and culture to shape the network society.

The above features of Korea's information revolution provide some important clues for thinking about its future, but they may not tell the whole story. After all, few people who experienced Korea in the 1970s or even 1980 would have imagined that the nation could develop as rapidly as it has.

Korea's network-centric approach

Korea's digital development also placed strong emphasis on network infrastructure. From 1980 onward, the nation's leaders saw telecommunications and the digitization of network infrastructure as the key to the future growth and success of its electronics industry as a whole. In important respects, Korea pursued network-centric digital development, with a strong, continuing emphasis on the manufacture and export of the hardware used to build ever more modern digital networks.

The Korean experience over nearly four decades shows how much can be accomplished by focusing on building and advancing network infrastructure. The construction of fixed and mobile digital networks is a long-term, expensive project. Given continued advances in network technologies, it is literally a never-ending project.

For Korea, the project originated out of a crisis situation in which the Ministry of Communications had been unable to provide adequate telephone service. The revolutionary advances of the 1980s in electronic switching, the semiconductor industry and completion of the PSTN were just the beginning. They were followed by network innovations in each of the following decades, right up to the present. As noted by a World Bank study, "The key feature that distinguishes Korea's deregulation and competition policy in the telecommunications services sector from other countries was its reliance on facility-based competition."[3] Such competition results when new entrants into the sector build their own facilities to provide services, as opposed to service-based competition, where the entrant uses the facilities of the incumbent. Korea is one of the few countries that has multiple operators in all markets within the telecommunications services sector.

The Korean experience demonstrated the strength of human networks that included government organizations, private companies, academic and research organizations and the public. In the 1980s the government played a major role, but one of its initiatives was to begin privatization of telecommunications, starting with the creation of the KTA, predecessor of today's KT. With that action, the government created an industry where none had existed before, and the role of Korea's private telecommunications and electronics companies continued to increase over the decades covered by this study.

It is understandable if policymakers, corporate executives and citizens in developing countries might find Korea's network-centric approach to digital development discouraging. After all, disparities in the coverage and characteristics of digital networks among nations are one thing that defines the concept of "digital divide." However, given the continuous and cumulative nature of network technology and standards development, it seems imperative that all nations pay attention to digital network infrastructure. An important related consideration is that all of the following emerging technologies depend on extensive, reliable and fast digital networks:

- Augmented reality
- Virtual reality
- Robotics
- Artificial intelligence
- Blockchain

Take robotics, for example. To grasp the reliance of robots on networks, one only needs to ask what use is a robot that cannot communicate with a human controller or with other machines. The point of this discussion is simply that future development of digital technologies to address the SDGs or disaster risk reduction will depend heavily on the capabilities of future digital networks.

Infrastructure for sustainability

Beginning in 2008 Korea's digital development took on an added dimension with the policy, legal and institutional acknowledgment that the ICT sector would play an important role in achieving green growth and sustainable development. The shift toward green growth as a matter of national policy would affect the entire Korean economy, including its leading ICT sector.

Korea's dramatic shift from a brown growth to green growth strategy in 2008 had an effect on all sectors of the economy, but most especially the ICT sector. As argued in this book, it is difficult to conceive of achieving sustainability without employing the capabilities afforded by evolving digital networks and related technologies. More generally, infrastructure is central to quality of life in the 21st century. Sustainable development goal (SDG) number 9 explicitly refers to infrastructure, calling for building resilient infrastructure, promoting inclusive and sustainable industrialization and fostering innovation. Furthermore as argued persuasively in a post on the International Institute for Sustainable Development's blog, infrastructure development is implied in many of the goals as follows:

SDG 1 – end poverty in all its forms everywhere – the targets relate to access to basic services, building resilience and reducing vulnerability to climate-related extreme events, and other economic, social and environmental shocks. Good infrastructure is needed to provide this resilience, as well as for public service delivery, such as education, healthcare or access to water and energy.

SDG 2 – end hunger, achieve food security and improved nutrition and promote sustainable agriculture – the targets refer to an increase in investment for rural infrastructure, which illustrates the importance of infrastructure investment, not only in urban but also in rural areas.

SDG 3 – ensure healthy lives and promote well-being for all at all ages – target 3.8 focuses on access to quality essential health-care services for which the development of health centers and hospitals in urban and rural areas will be essential.

SDG 4 – ensure inclusive and equitable quality education and promote lifelong learning opportunities – target 4.a demands the construction and upgrading of learning facilities.

SDG 5 – achieve gender equality and empower all women and girls – target 5.4 points at the need for provision of public services and infrastructure for social protection of unpaid care and domestic work.

SDG 6 – ensure availability and sustainable management of water and sanitation for all – this goal and the underlying targets focus on availability, access, and sustainable water management, all which require carefully planned infrastructure projects.

SDG 7 – ensure access to affordable, reliable, sustainable and modern energy for all – targets 7a and 7b refer explicitly to the promotion of investment in and expansion of energy infrastructure.

SDG 11 – make cities and human settlements inclusive, safe, resilient and sustainable – targets relating to infrastructure planning or issues such as waste management, transportation, climate change mitigation and adaptation, and resource-efficiency, require sustainable infrastructure development to reach this goal.

SDG 12 – ensure sustainable consumption and production patterns – target 12.7 refers to the implementation of sustainable procurement practices and policies that will have to be reflected in the procurement of infrastructure projects as well.

SDG 13 – take urgent action to combat climate change and its impacts – this goal implies that infrastructure projects have to be structured in a way that helps on the mitigation and adaptation front, as well as being explicitly developed to protect the poor and vulnerable groups of the effects of climate change.

SDG 17 – the means of implementation of the SDGs and post-2015 agenda – the targets refer among others to multi-stakeholder partnerships. Public-private partnerships (PPPs) will become increasingly important as a way of delivering infrastructure.[4]

As of 2018, Korea had taken some bold steps to establish the legal, political and institutional structure for a transition from brown growth to green growth. Furthermore, its robust, world-leading broadband network infrastructure helped position it to become a world leader in smart grids and next generation energy infrastructure. However, the nation's efforts thus far had not yielded significant reductions in greenhouse gas emissions.

Education, research and development

Education is the basic process of the information age. As such, it is a prerequisite for research and development, leadership and all the other key factors in Korea's digital development.

Education played a prominent role in Korea's transformation in several different ways. First, the nation built up its infrastructure for formal education, a process that by the 1980s and 1990s came to focus on university-level education. Second, in building the nation's first digital networks, it gave priority to first connecting schools and universities. Third, Korea strengthened its capacity for vocational education along the way. ICT is skills demanding and an essential tool for building a knowledge economy. A fourth component of South Korea's approach to education is study abroad. No other nation in the world sends as many students overseas for study, and considered on a per capita basis, Korea's lead is near insurmountable. A fifth contribution of education to Korea's information revolution was in creation of citizen awareness and demand. The continuing decades-long government, media, citizen and private sector campaign to promote the information society and raise public awareness of information culture, is itself one large educational undertaking. Today citizen awareness of the information society and its significance is probably higher in South Korea than in any other nation on earth. In the creation of citizen awareness and demand, Korea has set an example that many other nations might follow.

Another dimension of education's contribution to digital development in Korea is the integral role of research and development. The nation's research and development efforts in the 1980s focused on achieving an indigenous technological capability, initially in electronic switching systems and semiconductors. That approach to R&D proved to be nothing short of revolutionary.

Recognizing the continuing educational challenge, both at home and globally, in July of 2009, three government agencies were consolidated into the Korea Internet and Security Agency (KISA). The old agencies included the Korea Information

Security Agency, the National Internet Development Agency of Korea and the Korea IT International Cooperation Agency.[5]

The Korea International Cooperation Agency (KOICA) remains one of the most important government organizations when it comes to international educational aid, especially for developing countries. KOICA supports a very active program of aid on the role of ICT in development.[6]

Long-term planning and financing

As documented in Chapters 2 and 3 Korea's digital development originated in the early 1980s with the *Long-Term Plan to Foster the Electronics Sector*. Remarkably, the national commitment to building and improving digital networks continued over all subsequent presidential administrations representing diverse political views. The political commitment to build the nation's networks and ICT sector remained strong over time, even given considerable political and economic vicissitudes.

In addition to political and policy consistency supporting digital development, Korea also planned for the great costs involved. The 1980–1981 *Long-Term Plan to Foster the Electronics Sector*, specified that corporate participants in the telecommunications business be required to invest a specified percentage of their profits in a development fund to be used for R&D. Consequently, the PSTN project, Korea's first fully digital network completed in June of 1987, involved systematic reinvestment of a percentage (3 percent) of profits over a period of years into research and development, a percentage that was not reduced until 2002. Traditionally, spectrum auctions and licenses have been very lucrative for governments, and the funds generated are absorbed in the government's general budget for funding any projects as the government sees fit. In contrast, the Korean government recognized that, as a way to help Korea become a world leader in ICTs, these funds could be strategically reinvested in the telecommunications sector.[7] This was the first time a government had made such a long-term commitment to financing R&D and it helps account for the success of the TDX electronic switching project, the successful commercialization of CDMA and many others to follow.

Leadership and public-private partnership

Over the 30 years covered by this study, Korea's model for technological innovation changed in important ways. In the 1980s the state and its lead research institute, ETRI, exerted strong, top-down leadership as illustrated by the landmark TDX project. Over the ensuing decades, the private sector, not only in Korea but globally became more influential in funding and guiding telecoms R&D, and so the whole enterprise became more decentralized.

Far from retreating to the sidelines, the government assumed a role that we have likened to that of an orchestra conductor. In that role the government continues to this day to play a major leadership role, but does so in close collaboration with the private sector, academia and the public. If there is a lesson to be learned here from the Korean experience it is that inclusion of all the major stakeholders in ICT sector policy can lead to success that has broad public support.

The Korean experience underscores the reality that modern digital networks are large and costly construction projects. Part of the reason for this nation's success is that it invested significantly and over a long period of time in building the required infrastructure. As the experience of the United States and other countries in recent decades indicates, if private telecommunications companies are left alone, they may or may not invest adequately in a national infrastructure.

In the case of both broadband internet and mobile communications, the Korean government orchestrated the players, rather than letting the pure market oligarchy of large industry conglomerates rule. It attempted to level the playing field for competition, but at prices for affordability to assure public acceptance. The Korean government also showed a flexibility and willingness to change the road map as it unfolded in response to supply and demand.[8]

Our analysis suggests that Korea's approach to ICT sector policy became progressively more decentralized over the three decades covered by our study. One important aspect of this trend toward decentralization was the government's own effort to liberalize the telecommunications sector, in order that it could develop more fully in line with global trends.

As a recent report by the Information Technology and Innovation Foundation on *Explaining International Broadband Leadership* notes, "Demand-side policies matter." It cites South Korea's Agency for Digital Opportunity and Promotion (KADO), which merged into the NIA in 2009 and the sole mission of which is to promote digital literacy, access to computers, including training programs to let people buy computers through low-cost installment programs.[9]

Cybersecurity

In November of 1988 the young internet experienced its very first worm attack. At that time, there were only about 60,000 computers attached to the internet, and most of them were mainframes, minicomputers and professional workstations in government offices, universities or research centers. Within the span of one day, 5 to 10 percent of all internet-connected computers were compromised by the worm.[10]

That first worm attack was successful because of the open structure of the internet and the fact that computers back then were unsecured. This is what Zittrain refers to as the generative character of the internet. He argues that it is now in jeopardy because of developments since 1988. Unlike then, there is now a business model for bad code, resulting in a massive increase in computer viruses, malware and botnets. Leading anti-virus companies have begun to publicly express doubts about whether they will be able to withstand the growing onslaught of computer viruses.[11]

Today many individuals and organizations are responding to the increase in viruses, malware and spam by turning to cloud computing and the use of such electronic appliances as the Apple iPhone. The underlying question is whether this response will ultimately forestall internet failures caused by bad code.[12]

Since the first edition of this book was published in 2011, the global discussion of cybersecurity has taken on new urgency. In July of 2018 the U.S. Department of Justice indicted 12 Russians for hacking Democratic National Committee e-mails during the 2016 presidential election campaign in the U.S.[13] Additional revelations in 2018 suggested that "fake news" and Facebook posts by Russian hackers had reached millions of Americans during the 2016 presidential campaign. Such phenomena were not limited to the U.S. or Korea, but extended to other nations in Europe and around the world, as would be expected given the inherently global reach of the internet.

Over the years, Korea has had its fair share of experience with viruses, worms, malware and botnets. The July 4, 2009, simultaneous cyber-attacks on websites in the United States and South Korea were a sharp reminder of the increasing challenge posed by the threat of cyber-warfare. More recently, as noted in Chapter 9, Korea ranked fourth in the world in the total number of lost or stolen data records since 2013 and second in the world after the U.S. in the ratio of data records stolen to population.[14]

Korea's future place in cyberspace will depend upon its ability, together with governments, corporations and citizens groups around the world, to successfully address these serious security threats. Without a secure internet, the healthy growth of cyberspace is jeopardized. A secure internet, in turn, requires a viable form of international governance. This is precisely where South Korea can play a constructive role in the years ahead.

iPhone shock and Korea's relative weakness in software and services

The long-delayed arrival of Apple's iPhone in the Korean market gave a shock to the entire ICT sector, which until that point thrived on the manufacture, sale and provision of services utilizing domestically manufactured CDMA handsets. As this study showed, both the telecommunications service providers and the handset manufacturers appeared baffled by the iPhone, which was really a small handheld computing and communications device.

The big shift represented by introduction of the iPhone into the global marketplace was toward the creation of a mobile apps ecosystem. In other words, the new ecosystem would revolve entirely around software solutions and the services they could offer consumers. This did not mesh well with Korea's heavy reliance on the manufacture of electronics, appliances and hardware for building networks, nor did it match the structure of the global ICT industry, in which more than three quarters of the market is accounted for by software and services.

Korea's future place in cyberspace

The story of Korea's digital development as told in this book originated in its "telecommunications revolution of the 1980s" and unfolded alongside the global digital revolution and three important secular trends. Those trends were urbanization, globalization and the rise of concern over environmental and planetary sustainability. Korea's digital development also stayed on course over four decades despite some major political and economic disruptions. The worries expressed in the 1980s by leading technocrats, including Dr. Oh Myung, that Korea might miss the information revolution were not realized. To the contrary Korea is now riding atop the wave of the digital revolution that will power the 21st century and beyond, as the lessons summarized in this chapter testify. What does the future hold?

The era is coming when well-made software and well-established venture companies will raise a country's GNP. It seems we are approaching an era in which all countries will be fighting for territory and position in cyberspace. Will Korea be able to meet this challenge appropriately?

As documented in this book, South Korea constructed the necessary hardware and infrastructure, becoming a world leader in broadband networks. However, it continues to lag behind in software and development of what we might call the information mind. A significant part of this problem involves language.

Computer and internet literacy is one part of the linguistic challenge. At the stage when Korea was approaching 10 million internet users and 20 million hand phones there was a delay in developing the capacity to use information. Until that point about half of South Korea's citizens were completely unable to deal with computers and could be classified as computer illiterate and internet illiterate. At that time the only practical use of computers for many people was simple web forms or playing games. Korea met that challenge successfully.

A second part of the language challenge has to do with the use of Korean versus English, Chinese or other languages of the internet. The creation of *hangul*, an indigenous alphabet, is considered to be a crowning achievement of Korean culture. It not only contributed to mass literacy in Korea, but also accelerated the diffusion of computers, mobile phones and other electronic devices that require keyboard input. Moreover, because of its alphabetic and scientific character, it facilitated the development of a thriving graphics industry in South Korea.

While it is natural that Koreans are most comfortable surfing the internet using their native language and alphabet, it is equally apparent that new applications, services and software must be developed in English and other dominant languages of the internet in order to succeed in the global marketplace.

In certain parts of cyberspace, South Korea has already established a strong presence. In the future, what will be the salient features of Korea's place in cyberspace?

We have addressed some of the answers above. Several things seem certain:

- The information and communications revolution that transformed South Korea is continuing apace.
- Education, innovation and technology change not only drive South Korea's economy but link it inextricably to nations all over the world.

- Korea's future information society will be increasingly multicultural, multi-lingual and global.
- Korea's place in cyberspace will depend far more on innovative software, applications and content than upon manufacturing of hardware for the under-lying networks.

In the short term, Korea appears poised to continue on the trajectory of network-centric, ICT-driven development. It appears likely to become the first nation in the world to have nationwide 5G and public safety networks, as well as commercial giga-internet service. These next generation networks will offer many possibilities to strengthen competitive sectors of Korean industry, based on big data, artificial intelligence, robotics, augmented reality, virtual reality and blockchain technology. Korea's leadership in digital networks offers growth opportunities for its industries and also makes the nation an important testbed for the world.

However, over the long term it appears certain that Korea's leadership in shaping the global information society will ultimately be most profoundly exerted in cyberspace. This seems like the appropriate space in which to envisage a society in which information flows freely among the many cultures of the world, yet with respect for diverse cultural traditions. In the future global information society, one clear challenge for Korea will be to continue the transformation from a monolingual, relatively homogeneous culture into a multilingual, multicultural and outward-looking nation. Indeed, the imperative of environmental and planetary sustainability would seem to demand such a development trajectory. No nation, let alone South Korea, can tackle such problems alone. Indeed, the World Bank President Kim Jim Yong had it right: "Lessons for Korea are lessons for the world."

Notes

1 John Perry Barlow. (1996, February). *A declaration of the independence of cyberspace*. Retrieved from http://homes.eff.org/~barlow/Declaration-Final.html
2 From course materials, Gary Stringer. (2006–2008). *Conceptual Issues in Cyberspace*. University of Exeter. Retrieved from http://services.exeter.ac.uk/cmit/modules/cyberspace/webct/ch-philosophy.html#id690561
3 Suh, Joonghae and Derek H. C. Chen. (2007). *Korea as a knowledge economy: Evolutionary process and lessons learned*. Korea Development Institute and The World Bank Institute, p. 85.
4 Casier, L. (2015, September 9). *Why infrastructure is key to the success of the SDGs*. Retrieved February 17, 2017, from IISD Blog: www.iisd.org/blog/why-infrastructure-key-success-sdgs
5 Retrieved July 20, 2010, from www.nida.or.kr/kisa/eng/english_ver.html
6 Retrieved July 20, 2010, from www.koica.go.kr/english/aid/ict/index.html
7 Forge, Simon and Erik Bohlin. (2008). Managed innovation in Korea in telecommunications – moving towards 4G mobile at a national level. *Telematics and Information*, 25, p. 304.
8 Forge, Simon and Erik Bohlin. (2008). Managed innovation in Korea in telecommunications – moving towards 4G mobile at a national level. *Telematics and Information*, 25, p. 305.

9 Atkinson, Robert D., Daniel K. Correa and Julie A. Hedlund. (2008, May). *Explaining international broadband leadership*. Washington, DC: The Information Technology and Innovation Foundation, p. ix.

10 Zittrain, Jonathan. (2008). *The future of the internet and how to stop it*. New Haven, CT: Yale University Press, p. 36.

11 Zittrain, Jonathan. (2008). *The future of the internet and how to stop it*. New Haven, CT: Yale University Press, pp. 46–47.

12 Zittrain, Jonathan. (2008). *The future of the internet and how to stop it*. New Haven, CT: Yale University Press, p. 101.

13 Swaine, Jon and Andrew Roth. (2018, July 14). US indicts 12 Russians for hacking DNC e-mails during 2016 election. *The Guardian*. Retrieved December 1, 2018, from www.theguardian.com/us-news/2018/jul/13/russia-indictments-latest-news-hacking-dnc-charges-trump-department-justice-rod-rosenstein

14 Sobers, Rob. (2018, July 16). The world in data breaches. *Varonis*. Retrieved November 29, 2018, from www.varonis.com/blog/the-world-in-data-breaches/

Dr. Myung Oh's education and career

Education

1958	Graduated from Kyunggi High School
1962	Graduated from Korea Military Academy
1966	B.S. in Electrical Engineering, College of Engineering, Seoul National University
1972	Ph.D. in Electrical Engineering, The State University of New York at Stony Brook
1997	Honorary doctoral degree in humanities from the State University of New York at Stony Brook
1999	Honorary doctoral degree in business administration from Wonkwang University in Korea
2009	Honorary doctoral degree from Universidad Nacional de Asuncion of Paraguay

Career

1966–1979	Professor of Electrical Engineering in Korea Military Academy
1979–1980	Research Fellow of the Agency for Defense Development
1980–1981	Presidential Secretary for Economy and Science
1981–1987	Vice Minister of Communications
1987–1988	Minister of Communications
1989–1993	Chairman of the Taejon International Expo Organizing Committee
1992–1993	Chairman of Korea Institute of Nuclear Safety
1993–1993	Commissioner of KBO (Korea Baseball Organization)
1993–1994	Minister of Transportation
1994–1995	Minister of Construction and Transportation
1996–1999	Chairman of DACOM
1996–2001	President/Chairman of *Dong-a Daily* newspaper
2002–2003	President of Ajou University
2003–2004	Minister of Science and Technology
2004–2006	Deputy Prime Minister and Minister of Science and Technology
2006–2010	President of Konkuk University

2010–2012	Chairman of Woongjin Energy and Polysilicon
2010–2013	Chairman, Korea Advanced Institute of Science and Technology (KAIST)
2010–2012	Chairman of National Cancer Center
2013–2014	Chairman of DB Hitek
2016–present	Honorary President, State University of New York in Korea

Index

Note: Page numbers in *italics* indicate figures; page numbers in **bold** indicate tables.

Printed in the United States
By Bookmasters